Re

Not for resale

Field·Study

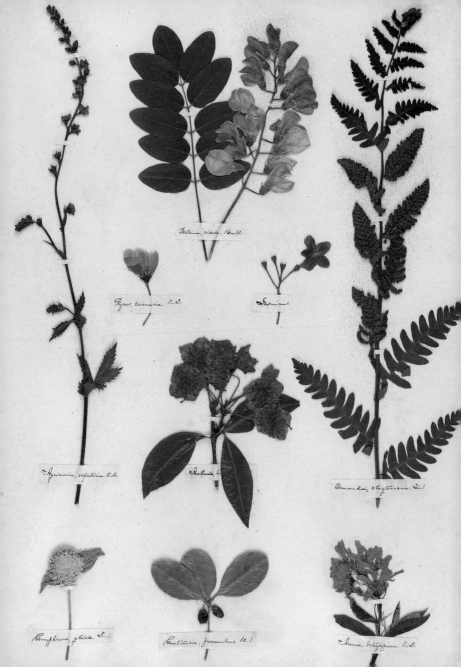

Robinia violacea. 16=10.

Pyrus coronaria. 11=5.

Jasminum.

Agrimonia, eupatoria. 11=5.

Kalmia la.

Osmunda, claytoniana. 2=1

Osmunda, claytoniana. 2=1

Gomphrena, globosa. 5=1.

Pimelea, procumbens. 12=1.

Ixora, stricta, aspium. 11=5.

Meditations on a Year
at the Herbarium

Field
Study

‹‹‹○›››

Helen Humphreys

ECW PRESS

This book was researched and written on the traditional territory of the Anishinaabe and Haudenosaunee peoples. I am grateful for their long and vital relationship with the plant life that inhabits this region.

For KW

CONTENTS

"Herbarium specimens are the physical
vouchers for the world as it was."

Deb Metsger
CURATOR OF THE ROYAL ONTARIO MUSEUM
GREEN PLANT HERBARIUM

‹‹‹‹‹‹●››››››

"There are days we live
as if death were nowhere
in the background; from joy
to joy to joy, from wing to wing,
from blossom to blossom to
impossible blossom, to sweet impossible blossom."

Li-Young Lee
"FROM BLOSSOMS"

Introduction

⟨⟨⟨⟨⟨⟨⟨⟨⟨ • ⟩⟩⟩⟩⟩⟩⟩⟩⟩

The autumn leaves floating down over the field look like brightly coloured birds falling to earth. We have just left the pine wood and are on the path that my walking companion calls Carnage Alley, because there are often feathers, or blood, or bits of dead animal on this route. The victims of coyotes, perhaps, or the owls that hunt above this field at dusk. Easier to catch something in the open than in the tangled wrack of forest trees, and even now there is a northern harrier skimming the tops of the asters and milkweed. We always consider it lucky to see the harrier, so we stop to watch its low, silent glide. It seems otherworldly, an owl's head on a hawk's body, the elegant drift of its hunger.

This place of woods and meadow and marsh is paradise. My paradise. Where I walk every day, all through the seasons. It always seems to be teeming with wildlife and plant life, but things have changed even in the handful of years I have been coming here. Now there are deer ticks on all the forest paths and in the open fields. The toxic wild parsnip is creeping through the meadows, and an invasive feathery reed, phragmites, is choking out the wetlands. The bobolinks and meadowlarks, who used to be plentiful every summer, are now virtually non-existent.

Habitat loss, pollution, climate change, human overpopulation and encroachment — these are some of the main reasons for the decline and changes to ecosystems. Much of the damage is irreversible, and the prognosis for the future is grim. And yet, I believe there is still a profound need within human beings to connect to the natural world.

How to reconcile these two things?

Increasingly, this morning walk I take through the woods and fields with my dog and a friend has become crucial to my physical and mental health. Without it, I have difficulty handling all the stresses of this world, and all the losses that have occurred in my own life.

I am interested in exploring this relationship, to write from a place that doesn't look away from the environmental changes wrought by humankind and that also celebrates the connections that still exist between people and nature.

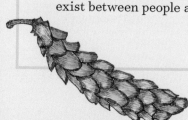

To do this, I have chosen to concentrate on the phenomenon of the herbarium. These libraries of dried plant specimens — some hundreds of years old — seem the perfect crucible in which to examine the intersection of human beings and the natural world through time. Each herbarium specimen is mounted on a sheet of paper with a label affixed by the collector, providing details of the plant and the location where it was found, but also including information about the person who preserved the plant. In this way the herbarium becomes a place, a landscape if you will, where the experience of people connecting with nature is revealed. I cannot think of another place where it is possible to look into the past and see the moment an orchid was plucked from the forest floor or a willow frond was cut from a branch. A visit to the herbarium is an exquisite kind of time travel. And by learning more about the intersection of people and nature in the past, I hope to gain some understanding of where we can go from here.

⟨⟨ ◊ ⟩⟩

Winter

Mimulus ringens. L. D.

‹ ‹ ‹ ‹ ‹ ● › › › › ›

The Herbarium

specimen

ALLEGHENY MONKEYFLOWER
(*Mimulus ringens*)
Emily Dickinson

The road to my particular herbarium — the Fowler Herbarium — winds through forest, twisting like a river, each turn revealing something new and surprising: a rafter of wild turkeys in the woods; deer browsing on the underbrush; a glittering, icy pond fringed with rushes; and, once, a fox nose down, snouting the snowy furrows of a winter field.

This herbarium originated in the late 1800s under the auspices of the Reverend James Fowler and was once housed in Queen's University in Kingston, Ontario. It now has its own building at the site of the university's biological station on the edge of Opinicon Lake: a climate-controlled, metal-clad structure that houses over 140,000 dried plant specimens. I mean to look at each and every one of them.

A couple of hundred years ago, it seems as if literally everyone picked and pressed flowers and plants and made a herbarium. Thoreau had one, as did Emily Dickinson. Botanizing was a popular settler pastime in the nineteenth century, both on a professional and amateur level, with lots of cross-pollination between the two groups. One aspect of colonization was a feverish desire by the incomers to catalogue the flora and fauna of North America, and since the tools

required for botanizing were few —
a notebook and pencil, magnifying
glass, and specimen bag — it became
available to rich and poor alike.
The abundance of wilderness and
the minimal equipment required
to explore it was coupled with the
notion in the mid- to late 1800s of
self-improvement through acquired
knowledge. Many people who did
not have a formal education were
motivated to learn more about the
fields and forests that surrounded
their towns and villages as a way
to better themselves, and in doing
so, perhaps better their situation
in the larger world. Unfortunately,
many plants, including the familiar
Lady's Slipper Orchid (*Cypripedium
calceolus*) were driven to the edge
of extinction by enthusiastic plant
hunters and collectors.

LADY'S SLIPPER ORCHID
(*Cypripedium calceolus*)
Emily Dickinson

Before field guides, plant specimens were mailed to other botanizers for identification. Many of these experts were men and women with no formal scientific training, people who were simply passionate about a particular species of plant and had become an authority on it. In the twentieth century, this amateur-professional alliance ceded to a model of "science first," and the opinions of amateur botanists — many of whom had become leading experts in the field — were no longer sought out by the qualified scientists. When these amateur experts died, their collections were distributed among the many institutional herbaria, or destroyed. As a result, the present-day herbaria housed in universities and museums are composed of specimens from a multitude of both amateur and professional collectors, from all over the world. The herbarium has become a memorial to a kind of democracy that no longer exists in the same way in the scientific world, even with the advent of citizen science.

GERANIUM
(*Pelargonium*)
Emily Dickinson

Within the Fowler Herbarium the plants are organized sequentially, following a system of evolutionary biology — an order that begins with the oldest plant group, ferns. The specimens, separated into genus and then species, are contained within file folders, and the folders are stored flat and stacked on top of one another within metal cabinets.

Evolutionary biology is one way to organize a herbarium, but there are other methods, and one herbarium is not necessarily categorized in the same manner as another. The collection at Kew in England, for example, which numbers seven million specimens, is arranged by plant family, then genus, and then species. Within a species, the plants are then arranged according to their geographical location. At the Fowler Herbarium, the diverse places where the plants were originally found are mixed together and identified by the colour of file folders — blue for local origin, pink for the U.S., yellow for Europe, green for Greenland. The system put in place when a herbarium began is usually the system that is still being used by the institution where the collection is stored. While there may be a universal cataloguing system for book libraries, there is no such corresponding thing for plant libraries.

Before I start with the specimen cabinets, I look through
the collecting notebooks of Roland Beschel, an Austrian
botanist and lichenologist who took over the neglected
Fowler Herbarium in 1959 and was responsible for
renewing the collection, adding tens of thousands of
specimens and moving it to a location within Queen's
University. Beschel died suddenly in 1971 at the age of
forty-two and the directorship passed to Adele Crowder,
a specialist in bogs. Since Crowder's retirement, the
herbarium has been under the charge of Acting Curator
Adriana Lopez Villalobos, who is continuing the work of
digitizing the collection begun by Crowder.

Beschel's field books are a lot like my own notebooks,
in that they contain a multitude of things, not all of them
related to the task at hand. As well as his location notes
for the various specimens he was collecting, Roland
Beschel also put in personal observations, or drew
small sketches of a leaf or plant. He sometimes pressed
a flower or a frond of grass between the pages of the
field books, or included a list of expenses for a collecting
excursion. (On one such expense list the items include
cigarettes, sandwiches, a stay at the YMCA, and a present
for someone back home.[1]) He also noted what was around

him while he was on his field trips, talking about the first lilacs blooming, or stating that the sight of a young moose crossing a tidal bore in Moncton was "Not so impressive, but still surprising."[2]

What happens when working in the field is that the field itself begins to assert itself. In Beschel's notebooks there is evidence of his having been outside when he was taking his notes — a squished mosquito on one of the pages, raindrops smudging the ink from his pen, a stray piece of grass caught in the crease of the spine. Being outdoors doing fieldwork meant that Beschel was exposed to whatever was taking place there.

It also meant that he noticed other life forms that were in the vicinity of his study area. He wrote small lists of the birds he saw when he was out in the woods and remarked on sightings in previous years. He once saw a loon and her babies at a spot on the lake where he had gone to look for plant life, and he forever afterwards called the spot "Loon Islet."[3]

Roland Beschel was a good observer, not just of the plants he was interested in recording and collecting but also of the larger natural world. Reading his notebooks, I come to appreciate his sensibilities and the bits of his personality that are on display in some of his entries.

He is dedicated, curious, and humane, and I grow fond
of him as I read through all twenty-four of his notebooks.

He described the sight of fifteen nighthawks
flying together and said that when they were diving,
"the feathers rattle." When talking of meeting a fellow
botanist, he shrewdly observed his colleague's character
as being that of an artist — "Idealistic and yet frustrated"
— and went on to wonder, "Why does Fabrio not collab-
orate with him?" In a declaration of generosity, Beschel
wrote, "I will send him the liverworts from the Copper
Swamp."

What hadn't occurred to me before I began looking
through the herbarium notations and the collectors' field
books was that some scientists were better observers
than others, just as some artists are. It is not something
that can be fully taught, but is perhaps a product of
one's natural curiosity, and also maybe depends on how
well a person can put themselves aside — their desires
and ego — to focus on what is outside of and around
them. People can be trained, through art and science,
to be good observers, but to be a great observer perhaps
depends on something innate.

‹ ‹ ‹ • › › ›

Roland Beschel was a lichenologist. Lichens are a combination of a fungus and algae, and are the witnesses of the natural world, in that they do not change through the seasons. Everything else changes around them and they remain stable. They do not move into space the way vascular[4] plants do, but slowly accumulate and accrete, spreading outward not upward.

In the early days of North American settler science, botany was about taking an inventory of the plants that were here. Now the scientific focus has shifted to cataloguing the damage that the human world has perpetuated on the natural one, and lichens are being used to measure the effects of pollution on the environment.

25

LICHEN   Helen Humphreys

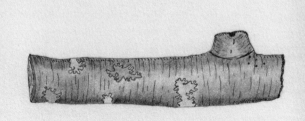

A lichen can be thousands of years old, growing in the spot it has always grown in and registering within its cells the changes that have occurred in the atmosphere over time.

If a lichen is to be a reliable witness to the years, it has to grow in a protected place that is itself allowed to be unchanging. Cemeteries and churches, caves and mineshafts are good environments for lichens and, subsequently, lichenologists. Roland Beschel regularly visited certain graveyards to monitor the lichens on the tombstones. He made excursions to the Arctic and Greenland as well, locations where there was little human interference with the landscape and where the lichens could flourish undisturbed.

〈 〈 〈 ● 〉 〉 〉

Adele Crowder, in a paper she wrote about the Fowler Herbarium collection, remarked on the "romantic tastes"[5] of the nineteenth-century botanists, who liked to collect moss and lichens from famous places — in Paris, for example, at the tomb of Abelard in the

LICHEN   Ernst Haeckel

Père Lachaise cemetery, or from the mossy cobbles in front of the church of the Madeleine.

The collectors' notebooks have shown me that the interchange between the plant matter being collected and the human being doing the collecting yields details of the location of a particular specimen, the personality of the collector, and of the larger environment where the specimen was found.

The personality of the collector is evident in the way they arrange the specimens on the page. Some glue the plant down as a clump, ripped straight from the earth, with dirt still clinging to the roots. Others spend time separating out each leaf and frond and arranging the specimen in a more deliberate way, often creating a page of startling beauty. Some collectors take and glue a single specimen, and others attach multiple examples on the same archival sheet of paper. In this way the act of specimen preservation becomes a sort of signature of the collector.

Because reliable images and information for plant identification are now available online, a herbarium does not perform the same taxonomic function it once did. These days the herbarium's plant specimens are used for extracting DNA to help decode the evolutionary

history of related plants, and as references for the range
of variance in a specific plant, as there are multiple
examples of each plant from different locations around
the world. The herbarium is also, therefore, an enormous
illustrative catalogue that shows "what is where."[6]

《 《 ◎ 》 》

My interest is in showing the intersection between
nature and people, between the plant and the collector.
This field study is not a comprehensive document of
every plant and collector, but is instead my response to
what I have found gathered and catalogued within those
white cabinets. I have treated the Fowler Herbarium
as a kind of wilderness, following particular threads
that catch my imagination and curiosity. I want to show
how exploring an archive can be a journey, as thrilling
as any taken in an actual landscape, and that no two
journeys are alike. My decision to use the herbarium as
a wilderness is my particular path. Another researcher
or writer would choose a different direction. There is no
right or wrong way to go about this kind of exploration.
It is a matter of choices made and the interest and intent
that informs those choices.

《 《 ◎ 》 》

‹ ‹ ‹ ‹ ‹ ‹ ● › › › › › ›

Ferns

specimen

CUPLET FERN
(*Dennstaedtia bipinnata*)
Fowler Herbarium

When I first look through the herbarium cabinets,
beginning, as the collection does, with Ferns, I am struck
by how the plants are layered on top of one another
in the files, with each large, metal cabinet becoming
a sort of mausoleum. Everything is dead — the plants
and the collectors themselves, who are preserved with
the briefest flicker of ink on their specimen labels.

All of these people and plants, who were placed nowhere near to one another in life, are now jammed up against one another in death.

〈 〈 〈 ○ 〉 〉 〉

Here is William Starling Sullivant, who became a botanist after watching a man collect plants in a field outside his house while he was eating dinner. Sullivant rushed outside, leaving his uneaten plate of food on the table, and then invited the stranger in to dine with him and explain his botanizing.

Now it is September 18, 1841, and Sullivant has written a long letter to accompany the specimen of Carolina mosquito fern (*Azolla caroliniana*) that he has just scooped from a flooded meadow in Columbus, Ohio. The letter begins with an ink drawing of a hand, with the index finger pointing to the body of the letter, which explains at length how Sullivant has been watching this particular patch of mosquito ferns since 1840 and now, a year later, they are finally displaying reproductive organs and so he has decided to capture them and has glued several examples onto paper for his private herbarium.

〈 〈 〈 ○ 〉 〉 〉

And here is a beech fern (*Phegopteris dryopteris*) collected by Sir Charles James Fox Bunbury on June 10, 1844, from Ambleside, Cumbria, in the northwest of England. The fern was picked from the base of Stockgill Force, a seventy-foot-high waterfall that so impressed John Keats when he saw it on a walking holiday in 1818, that he vowed to try to write poetry to honour it and said of the sight of it, "What astonishes me more than anything is the tone, the coloring, the slate, the stone, the moss, the rockweed; of, if I may say so, the intellect, the countenance of such places."[7]

The beech fern grew in great abundance at Stockgill Force in the 1800s, so Bunbury's finding it was not made difficult by scarcity.[8]

The Victorians had a passion for ferning, and those with the means often had their own greenhouses and grew their own ferns to study and draw. Ferns were a fad, a fascination with the "exotic" and bringing that perceived exoticism back to Britain. It was a botany of empire.

BEECH FERN
(*Phegopteris dryopteris*)
Fowler Herbarium

In Australia, Baron Ferdinand von Mueller's collecting
team of men and women were picking a drooping
filmy fern (*Hymenophyllum demissum*) in the Upper
Yarra Valley in Melbourne to add to his extensive
herbarium, which he called the Phytologic Museum of
Melbourne. Later it became the National Herbarium of
Victoria, when he was appointed government botanist
of Victoria in 1853. At the time of von Mueller's death
in 1896, it contained close to one million specimens and
can still be visited today. Von Mueller had relied, as
many of the European botanists had, on a vast network
of three thousand people to collect plants for him,
largely amateur botanists.[9] He appealed particularly
to women, putting an ad in the *Brisbane Courier* in
April 1872 that read in part: "What trouble would it
be to collect and preserve flowers, and enclose in an
envelope to their destination? How many ladies might
devote a few leisure hours to this pursuit? Every
contribution would be acknowledged, and not only that,
but the donor's name become a part and parcel of this
great undertaking."[10]

‹ ‹ ‹ ◊ › › ›

In Canada, in 1862, a missionary ship called the *Dayspring* was built in the New Glasgow shipyards in Nova Scotia with funds raised by the Presbyterian Church. It was a large ship of 120 tons, a three-masted schooner or "barque," bound for a group of islands in the South Pacific known as the New Hebrides, and now called Vanuatu. Once there, the missionaries encountered slavers who were coercing the islanders to board their ships, transporting them to work in the cotton plantations of Fiji and the sugar plantations of Australia. The missionaries from Nova Scotia were anti-slavery and the crew of the *Dayspring* tried to warn the islanders about the trafficking, while also persevering in their mission to convert them to Christianity.[11]

The *Dayspring* was captained by William A. Fraser, a former smuggler from Nova Scotia, described as "a bold fellow, and ready for anything yet thoroughly honest and upright."[12]

Fraser collected ferns while he was in the South Pacific, and the stag's head fern (*Lycopodium cernuum*) in the Fowler Herbarium was plucked from the jungle floor on the island of Aneityum in 1872. This small island, just 61.5 square miles, was reported to have once contained over a hundred different species of ferns.[13]

Botanical label from the Fowler Herbarium.

ANEITEUM FERNS.

Lycopodium cernuum, L

Collected by J. FRASER,

Capt. of the Missionary Ship "Dayspring."

1872

37

A year after the stag's head fern was picked by Captain Fraser, the *Dayspring* was wrecked beyond repair in the Aneityum harbour during a hurricane. Fraser decided to retire and returned to Nova Scotia.

‹‹‹ ◦ ››

In New Jersey in the 1870s, Mary Treat collected a small curlygrass fern (*Schizaea pusilla*). She was one of only four American woman botanists who were publishing before 1880, and who earned enough from her books and scientific articles to support herself; when she was forty-four, her husband left her for a younger woman.

Mary Treat was self-taught and well respected by professional botanists. She was in correspondence with both Charles Darwin and Asa Gray, the famous American botanist. She didn't believe in ranging far for her specimens, spending most of her time in her own garden, botanizing and writing about her observations there. She believed that observing a small place yielded better results than travelling over a large area: "The smallest area around the well-chosen home will furnish material to satisfy all thirst of knowledge through the longest life." [14]

CURLY-GRASS FERN
(*Schizaea pusilla*)
Asa Gray

‹ ‹ ‹ ◦ › › ›

In 1903, Alfred Brooker Klugh picked a cinnamon fern (*Osmundastrum cinnamomeum*) from a place in Ontario he called "Indian Pipe Swamp." This was his name for the place, so there is no knowing where it actually was. An additional note on his collector's label, in someone else's hand, guesses that the location was somewhere in Puslinch township, in Wellington County.

A.B. Klugh, as he is known on his collecting labels, came to Canada from England with his parents when he was twelve. He graduated in nature study from the Ontario Agricultural College, following with an MA from Queen's University and a PhD from Cornell University. He taught as an associate professor in the botany department at Queen's until his death at the age of fifty, when the car he was travelling in was hit by a CN train at a rail crossing in Kingston, Ontario.

Klugh picked this cinnamon fern for his own herbarium on July 16, 1903, when he was twenty-one years old. According to historical weather data, the day was cool for summer, with a mean temperature of just 14.2 degrees Celsius, clear and without any rain.[15]

The cinnamon fern is a large fern that grows in clumps in wetlands. It gets its name from a single upward frond, or pinna, that is cinnamon coloured. The rest of the fern is green.

VIRGINIA CHAIN FERN
(*Woodwardia verginica*)
Fowler Herbarium

Woodwardia verginica (L.) Sm.

5 Aug. 1973   4296   I.D.Macdonald/J.L.
           IDM   1973
     UTM 762316   44°32' N   76°30' W

Bedford Twp.,Frontenac Co.,Ontario.
o.7 km. S of W end of Arkon Lake
very wey edge of quaking bog matte and drai
  in fringe scrub zone of Nemopanthus,Acer a
    Larix.

It is not hard to imagine this fern, growing amid other ferns, in a swamp that was also filled with Indian pipes, a plant so pale and otherworldly that it is also called the ghost plant or corpse plant. (It is so white because it doesn't produce chlorophyll as green plants do.) Into this swamp, on this cool July day, strides the young botanist, Alfred Klugh, at twenty-one almost halfway through his brief life. And here is this moment — the fern reaching up, the hand reaching down. And then the note on the location, a place Klugh names himself, a place that now only exists here, on this scrap of archival paper, in this blue file folder.

41

〈〈〈 ❊ 〉〉〉

Paleobotanist Ralph Works Chaney also collected a fern specimen — a Virginia chain fern (*Woodwardia virginica*) — from a swamp. This swamp was on Hamlin Lake in Ludington, Michigan. It was a blueberry swamp, but as Chaney wrote on the specimen label, it had been "burned over" when he reached down to pick his fern on July 9, 1910. The fern still carries a scent — not scorch exactly, but something earthy and slightly sweet.

〈〈〈 ❊ 〉〉〉

42

Daniel Cady Eaton collected many ferns and, in 1880, wrote *The Ferns of North America*. Many of the collectors were knowledgeable enough, either on an amateur or professional level, to write books or field guides on their chosen plants, as a way to provide a service, and as a way of organizing their research and expertise.

《《《 ◦ 》》》

Edith Bolan Ogden lived in Orono, Maine, and was married to Eugene Cecil Ogden, who was to become one of the most renowned North American aerobiologists for his seminal research into airborne pollen dispersal. He taught botany at the University of Maine and later became the state botanist for New York.

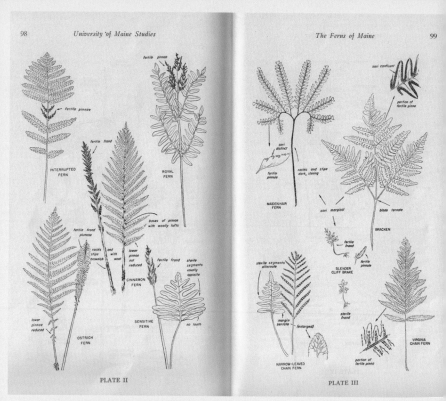

Page from Edith Bolan Ogden's *The Ferns of Maine*.

Edith Ogden worked with her husband collecting specimens for the University of Maine herbarium, and in 1948, at the age of thirty-seven, she published a guide to the ferns of Maine. She describes Klugh's cinnamon fern as being found "in boggy areas, wet pastures, river lowlands, basins, and around pools."[16]

While I was working my way through the Fern files at the herbarium, I had a recurring dream. In the dream, the dead people in my life were mingling with the living people and I could no longer tell which was which, who was alive and who was dead. It was as though I had forgotten this most fundamental distinction, and the dream was spent trying to sort out what was true and what was not.

‹ ‹ ‹ ◦ › › ›

The ferns in the specimen cabinets are no longer alive, but they still look as if they are, and though they are brittle, this dryness is not an indicator as to when a particular fern perished. A fern picked a hundred years ago is in much the same state of preservation as a fern picked ten years ago. The herbarium is a catalogue of dead plants, but perhaps it also tells us, equally, about what it is to be alive — that the dead and the living not only share the same space but are, in fact, equal.

‹ ‹ ‹ ◦ › › ›

Name *Pinus Stróbus, L.*

Common Name *White Pine*

Order *Coniferae*

Coll. *ALEX. H. D. RO*

‹ ‹ ‹ ‹ ‹ ● › › › › ›

Pines

specimen

WHITE PINE
(*Pinus strobus*)
Fowler Herbarium

The first thing I notice about the files in the Pines cabinets is the strong scent. Pines still smell piney, even after they have died. An American yew (*Taxus canadensis*) from 1902 still smells like itself, as does one from 1871.

One of the more interesting specimens in this section is a juniper (*Juniperus*). Bits of this tree are glued down in a haphazard lump, and there is an undated note below the plant instead of a proper label. The note reads: "The berries, which are bluish black, form an important article of commerce in Holland, where they are employed in the manufacture of Geneva, and impart to it that peculiar flavour which our distillers try to imitate by oil of turpentine. The wood is reddish and serves for veneering." The note is signed "Hooker."

Hooker is Joseph Dalton Hooker, the son of the famous English botanist William Jackson Hooker, who became the first full-time director of the Royal Botanic Gardens at Kew and started the herbarium there in 1853. The younger Hooker, after years of foreign collecting trips, eventually succeeded his father at Kew.

Joseph Dalton Hooker was one of the leading botanists of his day and specimens were frequently

COMMON JUNIPER
(*Juniperus communis*)
Fowler Herbarium

49

sent to him from all over the world for proper identification. He was a close friend of Charles Darwin and
was buried, like his father, at St. Anne's Church in Kew.
He died in 1911 at the age of ninety-four, having lived
through most of the nineteenth century.

The floral decoration on Hooker's memorial
medallion was drawn by his cousin Matilda Smith,
the official botanical artist at Kew for over forty years.
Hooker's daughter Harriet also became a botanical
artist — botany clearly being the family passion, and

business, and passing effortlessly through the DNA of one Hooker to the next.

The note from Joseph Dalton Hooker on the juniper specimen is personal. Although it is not addressed to anyone, it seems in response to something/someone, or meant for a display. There is the expectation of an audience for his musings on gin and veneering. But it is hard to know how, or why, this note ended up in the Fowler Herbarium.

‹ ‹ ‹ ◦ › › ›

I am starting to appreciate the randomness of the collection, a result of some specimens being deliberately collected for this particular herbarium, while others were donated or the consequence of a trade. Or, as perhaps was the case with the juniper specimen, part of a direct communication with a fellow botanist — amateur or professional.

‹ ‹ ‹ ◦ › › ›

William George Dore was an expert on grasses, but he collected many other plants that have ended up in the Fowler Herbarium. What sets him apart from other collectors is the amount of detail on his collecting labels. They often read like small stories of a particular plant.

For the eastern white pine (*Pinus strobus*) that Dore documented from a "Red sandstone island in lower Upper Beverley Lake" in 1966, he noted that the tree was a "natural bonsai," and he included a black-and-white photograph of the twisted tree growing out of a rock crevice. For another specimen, he included two photographs, the first showing his co-collector, retired forester Harold Zavitz, standing by the white pine he is preparing to sample, wearing white pants, a white shirt, a tie, and a fedora. One hand is in his pocket and the other holds on to a branch of the tree. Dore then quotes Zavitz saying, "I make it about 48, maybe 50 years old, likely the same age as the shrubby one." The "shrubby one" he is referring to is the "bonsai" tree in Dore's previous photo.

In the second photo, which accompanies a description so lengthy that it takes up most of a typewritten page and then continues overleaf, Harold Zavitz is again shown, this time in front of a pitch pine (*Pinus rigida*),

51

ONTARIO, CANADA

rigida (photo only)

ake, Leeds County, Ont.

ONTARIO, CANADA

Pinus rigida, PITCH PINE
Rear of Leeds and Lansdo[
Charleston Lake, Leeds County, Ont.

preparing to take a sample. His right hand reaches up into the branches and he is turned away from the camera. He has a collecting bag slung across his back. In the distance, but facing the camera, is a smiling young man in T-shirt and shorts, named on the label as Sam Vander Kloet, who was "studying the ecology of this tree localized in Leeds County." Zavitz, Dore notes in his label, was having difficulty with the young pine cones, which were closed when gathered and so he "had a hard time slicing off their side, exposing well-formed seeds."

It is a blow-by-blow of one collecting excursion, and there, glued beside the photographs and the long description of the pines, is the pine cone that Harold Zavitz had difficulty cutting in half, the once-fresh resin of its interior now dried to a kind of filmy varnish.

Dore worked at the Experimental Farm in Ottawa until his retirement in 1976, where he was a member of the Ottawa Field-Naturalists' Club and a founding member of the Canadian Botanical Association. He published a book called *Grasses of Ontario* in 1980.

53

PITCH PINE
(*Pinus rigida*)
Fowler Herbarium

Dore's descriptions and photographs record not just the details of the specimen trees but the inside of a particular moment. A warm, sunny July day. The photographer's coat thrown down casually on a rock near the base of the pitch pine. The two men, one distant and smiling, the other reaching up into the branches of the pine tree. And the third man who was there too that day, the one behind the camera, who would describe the moment in such detail, the one who set out to capture it all.

Dore lived to be eighty-four. Harold Zavitz died in 1978 at the age of eighty, and the young man in the photo in T-shirt and shorts, Sam Vander Kloet, who received his PhD in botany from Queen's went on to become a professor of biology at Acadia University in Nova Scotia and curator of the E.C. Smith Herbarium there. He was an expert in blueberries and died in the winter of 2011 while out on his evening walk in Wolfville, at the age of seventy-four.

‹ ‹ ‹ ◦ › › ›

Annie A. Boyd collected over four hundred specimens for the Fowler Herbarium over a two-year period, 1897 and 1898, when she was an undergraduate and then a master's student at Queen's.[17] The science program at the university had recently opened its classes to women students.

Annie Boyd collected plants in a range around Kingston that could be reached by bicycle. The white pine (*Pinus strobus*) that she added to the herbarium's collection was sampled on July 7, 1897, at Cartwright Point, a neighbourhood on the St. Lawrence River just to the east of Kingston, named for Richard Cartwright, a former United Empire Loyalist from Albany, New York, and one of the first British colonists to settle in Kingston in 1784.

After her time at Queen's, Annie Boyd took a job as a science teacher at Morrisburg Collegiate, remaining there for twenty-seven years.

Because Cartwright Point is close to where I live, I drive out to look at the white pines. It is not the summer day that Annie Boyd had when she bicycled across the causeway and up Barriefield Hill on her way to the Point, but is instead a cold April morning with a brisk wind and scraps of cloud moving fast across the pale blue sky.

Winter has just retreated from the landscape and the deciduous trees have yet to leaf, so it is easy to spot

55

WHITE PINE
(*Pinus strobus*)
Fowler Herbarium

Annie's white pines as they are the only green in the canopy and they tower twenty or thirty feet above the other trees. The Point has houses and cottages peppered along its spine, but they are inserted into the woods that cover the land, not carved out of it. So, the tree cover has probably remained fairly consistent from Annie Boyd's time to mine.

I drive the short length of the Point and back again, past the mossy slabs of limestone that rise out of last year's dead leaves. The trunks of the pines are too big to encircle with my arms, when I get out to try.

It is easier to see the scope of the pines from the road that runs across the top of Cartwright Point because it is on a rise and the land falls away below it. The white pines run in a feathery stutter through the whole Point, their massive green tops often touching, and all slanting about fifteen degrees to the east because the prevailing wind off Lake Ontario is westerly.

When Annie Boyd bicycled out here in 1897, she was probably collecting more specimens than just the white pine. She might have come for most of a day, leaning her bicycle up against one tree and then another, creeping through the woods in search of flowers and ferns. The white pine didn't have to be looked for; it was simply there in abundance, sighing above her head as she walked through the woods. The wind in the branches of the trees sounding exactly like her breath. Or mine.

‹‹‹ ❍ ››

Grasses

specimen

HAIRY WOOD SEDGE
(*Carex hirtifolia*)
Fowler Herbarium

Like a grassland itself, the grass specimens stretch on and on, cabinet after cabinet, file after file. I labour through so many of the same specimens — endless dried sedge and oat grass — until I tire of them and skip ahead to the Orchid cabinets at the end of the row. But the orchids, which I had imagined would be as delicate and pretty in death as they are in life, are mostly all the same shade of brown when pressed. All colour and delicacy is gone and, when flattened, their beautiful, showy extravagance is lost to a dried, pulpy mass.

So, I go back to Grasses with renewed appreciation for the fact that they keep their form and colour and do not look very different in death than they did in life. Sometimes they have been glued down on their pieces of archival paper in imitation of how they bend with the wind. I'm not sure if this is accidental or purposeful on the part of the collector, or the person doing the mounting, but I like the arc of 150-year-old rushes on the page, as though they were preserved in the exact moment of bowing to the breeze.

‹ ‹ ‹ ◊ › › ›

There are not only a multitude of dried grass specimens, but also a myriad of places where they were collected in the nineteenth century. They were taken from a "mossy runnel," and "an abandoned field," "a cold, mossy bog," and "Chalky pastures," "On the edge of a winter pond," and "Damp meadows." A.B. Klugh writes of his grass in 1909 that it is to be found in a "Damp opening in moist woods." In 1975, M.I. Heagy collected narrow small-reed at La Pérouse Bay in Manitoba "In silt bottom pool, in *Salix* [willow] dominated area," and then went on to note on his label that "Specimens do not fit either of the so-called species."

I find many labels that are cryptic. In 1932, in Surrey, England, C.E. Hubbard collects a type of brome grass from a "derelict hayfield. Very abundant near remains of old haystack." God help anyone who goes looking for that old haystack even a year later.

In September 1904, C.J. Haberer picked slender flatsedge in Utica, New York, "Along cold streams in pasture." In 1934, Ralph Kriebel collected cypressknee sedge from the "Knot-hole at the base of a tree in swamp" near Avoca, Indiana. A year later he collected lake sedge from a "Field bordering Half Moon Pond in Posey County." J.M. Greenman and two other collectors

picked sallow sedge in June 1909 from the "moist grassy places, near hotel," without ever saying what hotel it was, or where it was located. It makes me wonder if some of the collectors didn't want anyone to know where they had found their specimens, wanted to keep the location secret and particular to themselves.

《 《 ◊ 》 》

On the other side of the equation, Wilbur H. Duncan is more exact on his labels. When collecting giant cane in Georgia in 1969, he describes the location as being "In open low level area between two level streams 3 miles east of Athens," and then goes on to note that "Plants dead by 29 June 1969." In April 1955 he takes a specimen of bladder sedge "In moist area in small ravine in open deciduous woods on steep east-facing slopes on w side of Chestatee River, 7.5 mi w 8 degrees N. of Gainsville. Elev. 900′.″

And there are disagreements. In 1892 on the Miramichi River, an unknown collector wrote on the official Fowler Herbarium label that the plant they had taken was sterile sedge, then someone else crossed it out and wrote star sedge in the line below. That was crossed out by Fowler himself, who then put sterile sedge back in and wrote in thick pen strokes beside it "This is true, C. Sterilis!"

Botanical label from the Fowler Herbarium.

W.G. Dore, whose specialty was grasses, wrote another beautifully descriptive label for his specimen of skunk cabbage. He generously described the location as being "Two miles north of Prescott, Grenville County" and then offered more detailed directions: "(Go south from the side-road about half-mile E. of Domville highway, over the sand hills, past the deep forest on the sand ridges, and into the extensive cedar and ash swamp woods. The Skunk Cabbage is thick through this springy sodden land; not seen elsewhere in district despite the presence of suitable habitat; perhaps adventive.) Thuja woods on muck; abundant over several acres."

The vastness of the grassland at the herbarium is indicative of the popularity of grasses among botanists and collectors. For one thing, they have a direct link to agriculture and pasture land, and some of their study will be related to improving farming practices. Secondly, grasses are, literally, everywhere. A botanist does not need to take long excursions through remote woodlands, searching for a single rare, elusive plant. The grass collector can find any number of varieties growing in the ditches at the side of the road, or along the bank of the smallest stream. The ease with which it can be collected makes it a desirable specimen to collect.

‹ ‹ ‹ ◦ › › ›

Ægopodium podagraria, L.

Weehawken. June 17 and July 5, 1871.

This plant has also been found in the neighborhood of Brooklyn. It is probable that it comes in with the German immigration.

25.

Wm. H. Leggett.

Herb. Univ. Cant.

BRITISH ISLES

Glaucium flavum Crantz

Farnham Kent

near Paulsle England 1838.

in herb. C. J. Bunbury

PROVINCE OF NOVA SCOTIA

CAPE BRETON ISLAND.

Linum catharticum, L. Abundant in damp spots in an old field,

SYDNEY.

M. L. FERNALD.

August 20, 1902.

Botanical labels from the Fowler Herbarium.

In the middle of the herbarium grassland, I feel overwhelmed by the number of collectors and their individual stories.

Hewett Cottrell Watson was a shy botanist, phrenologist, and evolutionary theorist from nineteenth century Britain. He corresponded with Darwin and invented a geographical portioning of Britain for the purposes of biological cataloguing that is named after him (Watsonian vice-county system) and is still in use today.

George William Clinton was a New York lawyer, later a judge, and also the mayor of Buffalo from 1842 to 1843, and an advocate for the building of the Erie Canal. He was an ardent amateur botanist and had his own herbarium, engaging as well in the exchange of specimens with other botanists. On the 7th of November in 1807, he was collecting in a cemetery in Albany when he had a heart attack and perished, at the age of seventy-eight. He was buried holding the clover that had been in his hand when he died.

Heinrich "Henry" Karl Daniel Eggert emigrated from Prussia to St. Louis, Missouri, where he collected and sold grapevine seeds to vineyards in Europe. For a long time he couldn't make a complete living from this, so he delivered newspapers morning and evening for twenty years, eventually developing paralysis in his left arm from the weight of the paper bag. He shot himself in 1904, at the age of sixty-three, leaving a personal herbarium of over sixty thousand plants.

R.C. Hosie published *Native Trees of Canada*, a book that went through many editions and printings and was reinterpreted as an art book in 2010 by the writer and artist Leanne Shapton.

FEWFLOWER SPIKERUSH
(*Eleocharis quinqueflora
prev. pauciflora*)
Fowler Herbarium

The Reverend A.C. Waghorne was a Newfoundland Anglican missionary priest who was an avid collector and published numerous pamphlets on the flora of Newfoundland.

John Harbord Lewis was a British plant collector in the late 1800s with his own herbarium, who then developed a passion for microscopy, placing a newspaper ad offering his herbarium for sale: "Herbarium, British and foreign, what offers?"[18] Not long after that he had moved on to rocks, minerals, and rare books and offered his microscope slides for sale or exchange: "Micro-mounts and material, mosses and dried plants, all correctly named, in exchange for illustrated or rare books, and photographs."[19] Plant collecting, while a passion, could, it seems, cede to other passions, should the mood strike.

Mary Agnes Chase was a self-taught botanist and botanical illustrator, employed by the United States Department of Agriculture (USDA) and the Smithsonian Institute. She became a leading expert on grasses and in 1951 published her *Manual of Grasses of the United States*, a book still in use today. Chase was also a suffragette who campaigned for women's right to vote, was imprisoned for it and force-fed while on a hunger strike. She was a pacifist, prohibitionist, and socialist and received an honorary degree at the age of eighty-nine (her only university degree) from the University of Illinois.

Wilhelm Suksdorf was originally from Germany, settling with his family first in Iowa and then in Washington State, where they farmed a few hundred acres. Shy and a self-taught botanist, Suksdorf eventually became a plant collector for the pre-eminent American botanist Asa Gray, and amassed a private herbarium of thirty thousand specimens.[20] He had a habit of naming his collecting locations after what he found there, or by using the name ascribed to the place by the Indigenous Peoples who moved up and down the Columbia River Valley. As a result, it was almost impossible for anyone to find any of his collecting sites after his death. In 1942 an entire master's thesis was devoted to figuring out Suksdorf's secret geographic world.[21]

Perhaps Wilhelm Suksdorf did not want anyone else to visit his Falcon Valley or Elk Ridge. He had a private relationship with Jackrabbit Gulch and Rice Creek and wanted to keep it that way. His non-scientific naming of locations was exasperating to the botanists who followed him, as was his desire to "split" variations of the same plant into different species.[22] But naming somewhere because of what you have encountered there makes more sense than any other way of naming and is a way to tie oneself to place. It becomes an acknowledgement of both the Earth's creatures and one human's individual experience. And I, for one, would love to visit Suksdorf's Butterfly Lake, wherever it is.

< < < < < ● > > > > >

Lilies

specimen

CANADA LILY
(*Lilium canadense*)
Henry David Thoreau

The cataloguing and preservation of North American plants is a product of settler culture, as was the botanizing craze of the eighteenth and nineteenth centuries. Indigenous Peoples have their own, much longer, relationships with the plant life in their territories. While combing through the collection, I have been mindful of this, and have also been looking for any interaction between the two groups.

I finally find something in Lilies.

The label that accompanies a specimen of flowered wild garlic (*Allium tricoccum*) records the collecting date of the specimen as July 17, 1858, and the location as "Portage from Indian Village to Bark Lake." All of this — the name of the specimen and the location and date — is written in ink, as is the name of the collector, W.S.M. D'Urban. But below all that, running along the bottom edge of the blue notepaper that the specimen is mounted on, is a line written in pencil. It is a short sequence in Ojibwe. The sequence is split in two. On the left is the word or phrase "Chi-kwa-kwich" and to the right is a phonetic spelling of it. Beneath that line are the English words "not well to do," or perhaps "not well" and "to do." Also in pencil, but in a different hand, and written above the ink label, are the words "Sudbury, Ontario," as the location of Bark Lake. In brackets, in between the two versions of "Chi-kwa-kwich," is the translation of the Ojibwe words into English: "It makes a bad smell."[23]

In the 1840s there was increased mining and logging activity by incomers along the northern shore of Lake Huron, and the Ojibwe people who lived and hunted there were ceded by the Robinson-Huron Treaty in 1850 to an area bounded by the first waterfall along the Mississagi River to the first waterfall along Blind River (known by the Ojibwe as Penewobecong). The Mississagi River connects by portage to Bark Lake, and so the "Indian Village" noted on the wild garlic specimen label is probably the settlement that was just east of the first waterfall on the Mississagi River.[24]

It seems clear from the phonetic spelling of "Chi-kwa-kwich" that D'Urban wanted to learn to speak the Ojibwe word for wild garlic correctly, or was practising speaking it. From an article he wrote for the *Canadian Naturalist and Quarterly Journal of Science* in 1858, he notes that for the few hundred specimens he collected in Quebec and Ontario, he gave their Indigenous names as much importance as their English ones. The son of an Algonquin chief from a settlement on the Rouge River in Quebec helped him with translation.

Allium Ursinum.   *Ail des Ours.*

The man was named Chi-Chick, which D'Urban noted was pronounced "Shes-sheep."[25]

William Stewart Mitchell D'Urban was British. He had followed his grandfather, who was a colonial officer and general, first to South Africa and then to Canada, where, at the age of twenty-one, he had collected the two flowers of wild garlic I found in the Lilies cabinet. He continued to collect plants all of his life, but he ended up back in England as the curator of the Royal Albert Memorial Museum in Exeter. D'Urban's interests obviously went beyond plants as he later published a fully illustrated book on the birds of Devon in 1892.

While it is wonderfully informative to have D'Urban's notation on the specimen, and to read through his list of plants with their Indigenous names in the article he wrote for the naturalist journal, it is only the partial story. He provides this note at the bottom of the journal article: "This Catalogue was completed in the summer of 1859, and a copy containing much more elaborate notes than those above, which I transmitted for publication at the beginning of February last by the Steamer 'Hungarian,' was lost on board that unfortunate vessel."

75

WILD GARLIC
(*Allium ursinum*)
Joseph Pierre Redoute

The *Hungarian* was a transatlantic steamship, wrecked on February 19, 1860, in a storm on the west side of Cape Sable Island, Nova Scotia. All 205 people aboard perished, as well as the "more elaborate notes" of D'Urban and his translation of plant names with Chi-Chick.

‹ ‹ ‹ ◦ › › ›

Why does a person choose to collect one plant over another? Does the nature of the plant somehow complement the nature of the collector? What draws two things together?

In the case of orchids, the attraction is easy to fathom. Orchids are delicate and beautiful, often elusive and rare — showy and secretive, all at once. The finding of them often requires long treks through bogs and into jungles, and they are prized not just for their beauty but also for the effort required in collecting them.

‹ ‹ ‹ ◦ › › ›

BOAT ORCHID
(*Cymbidium*)
Fowler Herbarium

Robert Pantling was an orchid collector and illustrator, specializing in Himalayan orchids. He published, along with Sir George King, a monumental work on orchids, *The Orchids of the Sikkim-Himalayas*, which documented the 449 different species present in the region. Using local people to do the collecting helped with the work of finding and amassing so many of the plants. Pantling's Himalayan specimens are the largest number of orchids by a single collector in the Fowler Herbarium.

‹ ‹ ‹ ○ › › ›

Oakes Ames the botanist (not to be confused with his grandfather of the same name, who was a manufacturer and member of the House of Representatives for the State of Massachusetts) had an orchid herbarium of over 131,000 specimens. It eventually became part of the extensive Harvard University Herbaria. Ames had spent his career at Harvard, first in its botanic gardens, and later in the botanical museum. When he died in 1950, at the age of seventy-six, his large estate in Massachusetts became the Borderland State Park, a history and nature preserve of over 1,800 acres.

The checkered rattlesnake plantain (*Goodyera tesselata*)
in the Fowler Herbarium was part of his collection.
It is a member of the orchid family and was collected
in September 1901 in Holbrook, Massachusetts.

‹ ‹ ‹ ◦ › › ›

Nothing has changed since I skipped ahead to Orchids
from Grasses and noted that they were disappointing
in their sameness when presented as specimens. I can't
reconcile the rareness of an individual orchid with the
brown mush they invariably become when pressed and
mounted. For the first time it occurs to me that I might
prefer some plants above others, and the characteristic
I seem to favour is for a particular plant to retain its
shape and texture, to look the same whether it is alive
or 150 years dead.

‹ ‹ ‹ ◦ › › ›

part two

Spring

Salix discolor, Muhl.

‹ ‹ ‹ ‹ ‹ ● › › › › ›

Trees

PUSSY WILLOW
(*Salix discolor*)
Fowler Herbarium

I have now been at the Fowler Herbarium through an
entire and long winter. Spring brings rain and flooded
fields, the reckless roadway shear of the nesting robin.
The ice on the driveway has given way to potholes. There
are eastern bluebirds perched atop the little wooden
birdhouses on the lane in to the biological station.

The willow specimens are as beautiful as the grasses.
They retain their shape and go one better in that they
also retain their texture. A pussy willow (*Salix discolor*)
from 1906 is still soft. The leaves on a strand of shining
willow (*Salix lucida*) from the penitentiary farm in
Kingston collected on June 24, 1893, still look glossy,
126 years later.

Many of the willows are from the collection of
Michael Schuck Bebb, or M.S. Bebb as he is known on
his collector's labels. Bebb was an amateur botanist
who worked in a pension office; he devoted most of his
free time to his passion for willows. He became, in his
lifetime, and through much study and observation, the
leading willow expert in North America and Europe.

One of the things he did to aid his work on the genus
Salix was to plant a grove of willows on the farm he
owned. In an area of about two acres, and lying along
a creek, he had discovered a collection of wild willows.

To this he added willows sent to him as cuttings by other collectors, both in North America and Europe. The grouping was extensive. In May 1873, for example, he writes about the thousand cuttings sent to him from Dr. Hooker at Kew that are growing well.[1] The willow plantation became the basis for his large private herbarium, some of which is now in these file folders open before me on this cool May morning — a morning when Bebb's willows would have had their first pale spring leaves.

PUSSY WILLOW (*Salix discolor*) Henry David Thoreau

The willow grove tended by Michael Schuck Bebb was later destroyed by the farmer who purchased the land after Bebb had to relinquish it for financial reasons. The new farmer cut down all the willows to make pasture land for his cattle.[2]

‹‹‹◊›››

M.S. Bebb's mounted specimens reveal a distinctive style. He took great care to separate out each frond and leaf and the delicacy of the arrangement on the page often looks like art. He sometimes augmented the specimen with precise pencil drawings of a bud or leaf. When I see his name on a collecting label, I know that I am in for a treat. The care with which he looked after his willow grove is reflected in the conscientiousness with which he has prepared this mounted willow branch for my gaze.

WHITE WILLOW
(*Salix alba*)
Fowler Herbarium

On this very day, May 3, in 1901, John B. Flett was
collecting willows to represent the flora of Washington
in the Smithsonian herbarium. He cut a piece of
heartleaf willow (*Salix cordata*) in Tacoma, noting that
he found it "Growing in pools which dry up." He also
collected a specimen of Pacific willow (*Salix lasiandra*)
that was seven to nine feet high and found growing
in water. He notes on the bottom of the label that the
willow was "frequented by honey bees!!!"

I like his three exclamation points, which still show,
all these years later, his excitement. I like it all the
more when I find out that he was forty-two when he
penciled in those jubilant exclamation points.

John Flett was originally from the Orkney Islands
in Scotland. He worked in Washington State as a
biology and geology teacher for twenty years, before
becoming a park ranger in Mount Rainier National Park.
He wrote a guidebook on the flora of the park and was
an avid climber, summiting Mount Olympus in 1907. He
provided mountaineering advice to climbers who came
to the park and was also part of a mountain rescue
team there. A glacier in the park is named after him.

Flett was not a willow specialist like M.S. Bebb.
His willow specimens are probably here because of a

87

specimen trade, or from some other random exchange or accession, so it feels a little lucky to have stumbled upon his two specimens of willow and the exuberance of his exclamation marks on the exact day that I am here in the herbarium.

‹ ‹ ‹ ● › › ›

There are several large, old willows growing along the edge of the lake near where I live, and I go to visit them at the end of one of my herbarium days.

They are gigantic, their trunks twisted from the speed and energy of their growth, and so massive around that when I encircle them with my arms, I barely cover a third of them.

Their new leaves are pale green and slender. They look like a school of small fish swaying together from the branches and they make a hissing sound as they move in the wind. The downward aspect of them — how the leaves hang down from the ropey fronds, and the fronds dangle from the branches — is like water falling.

‹ ‹ ‹ ● › › ›

When I was eight, my parents rented a summer cottage on a lake in northern Ontario. A family we knew rented the adjacent cottage, and there was much interplay among the eight children in that family and the three children who were in mine.

The father of the Skinner family in the adjacent cottage was a geologist who was often away working at various northern sites for mining companies. He was a tall, strong man, and was what my mother called *outdoorsy*. He was a very different sort of father from my opera-loving one, and I who, even at the young age of eight, felt drawn to the natural world was drawn to him.

Part of this summer vacation was spent with him teaching the children who were beyond toddler stage how to right a capsized canoe and how to paddle solo. I felt lucky to be included in these lessons, although he did not give much quarter to the fact that he was teaching children rather than adults.

I managed to right the canoe adequately, but when it was time for my solo paddle, the wind was up and I was blown halfway down the lake and, being only eight, didn't have the strength or stamina for the return paddle. No one came to rescue me, so I had to get back to the cottage by using the vegetation at the edge of the lake to haul myself along. It took hours, but nothing was made of my heroic journey when I staggered up the beach towards the cottage at the end of the day.

Mr. Skinner was always busy with his brood of children, instructing them in one thing or another, but one afternoon he was alone and I was alone. I can't remember what inspired him to do it, but he paddled me out to an island in the middle of the lake and walked me over the island, holding my hand and pointing out the plants and telling me their common names. I can still remember the way the golden light of afternoon drifted down like smoke, and the fallen tree we climbed over, and I can recall the feeling I had when the mushrooms and berries along the forest path suddenly had names. It was a feeling of delicious expansiveness, that this world that I had always felt connected to was now fully in relation to me because I knew the names of the entities that resided in it. That feeling has never left me and I will always be grateful to John Skinner — who died of a heart attack while he was in his fifties, younger than I am now — for that life-changing afternoon.

《 《 《 ◦ 》 》 》

The next large category of plants in the herbarium is Oaks, but to get to the oaks, I have to go through the birches and the alders, just as though I was traversing a real forest.

The notable aspect of the white or paper birch (*Betula papyrifera*) is that it was once called the canoe birch, in reference to the canoes made by Indigenous Peoples with the bark of the tree.

Just as though I was walking along the forest floor, the first thing I find in the Oaks section is an acorn. It is from 1893 and is still as shiny and perfectly intact as though it had been collected yesterday. My favourite label writer, William Dore, has a written description of another acorn, this one from a chestnut oak (*Quercus muehlenbergii*) that he collected on August 27, 1965, from Massasauga Point, Ontario; he quotes a Mr. E.C. Coon, who was perhaps

OAK
(*Quercus*)
Helen Humphreys

on the collecting trip with him: "He says the nuts are sweet and edible." Then Dore writes a "Comment" on his label to give a bit of backstory for Massasauga Point: "[It] is all limestone pavement; the original forest was early cleared and under grazing has grown up to a dense woods of Juniperus virginiana mainly. This large Chestnut Oak may have been left at a lot-corner. The Point was also an important Indian stopping place at 'The Narrows' of the Bay of Quinte."

〈 〈 〈 ◦ 〉 〉 〉

Most of the plants I have looked at thus far were collected in May or June, when they first flowered. But with the oaks, the sampling happens later in the year so that the acorns might be gathered as well as the leaves. On August 14, 1881, a tanbark oak (*Quercus densiflora*) was sampled from five thousand feet up in Siskiyou County, California, by Cyrus Guernsey Pringle, an American botanist and pacifist. Pringle had begun his botany career running a fruit nursery and successfully cross-breeding potatoes. He also operated a sort of hospital for damaged flower bulbs: people would send him the bulbs and he would return them to health.

A lifelong Quaker who believed fervently in peace, Pringle was conscripted into the Civil War but refused to fight. As punishment he was staked out on the ground for hours with arms and legs outstretched. At the end of the second day of this, and weakened from it, he wrote in his diary, "This has been the happiest day of my life, to be privileged to fight the battle for universal peace."[3] He was only freed from this punishment — which would likely have led to his death — and from the army, when the head of the USDA at the time, Isaac Newton, petitioned President Lincoln on his behalf.

93

NORTHERN RED OAK   (*Quercus rubra*) Henry David Thoreau

After this experience, Pringle went into serious plant collecting, contributing over 500,000 plants during his lifetime to various herbaria throughout North America. He also went back to plant breeding, and he was named botanical collector for the American Museum of Natural History in 1880. He was one of the most important plant collectors in North America, particularly active on the Pacific Slope, where he found and documented this sample of tanbark oak.

‹ ‹ ‹ ◦ › › ›

Another tanbark oak, this one collected by Lincoln Constance in 1934, has on its label perhaps the most animal references of any I have found here. His specimen was collected from the South Fork Eel River in Humboldt County, California, with the location given as "Cow Creek fire-trail, from Bull Creek Road, North over the divide, into the headwaters of Bear Creek." These are place names that Wilhelm Suksdorf would have approved of.

‹ ‹ ‹ ◦ › › ›

A sample of a gray oak (*Quercus grisea*) was collected from its native habitat in New Mexico in 1881 by George R. Vasey, son of George S. Vasey who was the chief botanist for the USDA and in charge of its herbarium, housed in the Smithsonian Institute. Vasey Senior was an expert on grasses. Vasey Junior collected widely throughout North America, living first in Washington state and later in Alberta. He was known for his discovery of the pinkshell azalea, which was later named for him (*Rhododendron vaseyi*).

‹ ‹ ‹ ◦ › › ›

95

Continuing on through the virtual forest, I come to a dwarf chestnut oak (*Quercus prinoides*), collected by Elizabeth C. Allmendinger on the bank of the Huron River in Ann Arbor, Michigan, in 1893.

Elizabeth Allmendinger was a local amateur botanist and plant collector who had her own herbarium. She compiled a detailed list of plants found within four miles of Ann Arbor that numbered 848 species. This list and her herbarium were later donated to the University of Michigan.[4]

SWAMP SPANISH OAK
(*Quercus palustris*)
Henry David Thoreau

There is just as much to be said of the flora that is close to home, as of the flora that comes from long-distance collecting excursions. I don't think that Elizabeth Allmendinger's life was less exciting because she concentrated on a four-mile radius around her home. It was easier, and safer, for a woman in the nineteenth century to remain close to home when she was botanizing, so it is not a reflection of a woman collector's sense of adventure or lack of it when she collects in a small radius, but rather a comment on what was possible at the time given societal restrictions for women.

‹ ‹ ‹ ◊ › › ›

There is a sample of a red oak (*Quercus rubra*) taken in June of 1893 from the cemetery where my brother is buried. So, I decide to go and have a look and see if it, or any of its brethren, is still there.

The cemetery was designed to resemble the park-like setting of Père Lachaise cemetery in Paris and has winding paths, small ravines and ponds, and a great many trees. It covers ninety-one acres and was founded in 1850, so the red oak in the herbarium, if planted when the cemetery first opened, would have been over forty years old when it was sampled. Now it would be upward of 160 years, which, given that the average lifespan for this tree is 200 years, makes it possible that it is still alive.

It's a sunny day in May when I drive through the cemetery. The maples are mostly leafed out, and several limbs are down from a recent windstorm. A slow drive around the perimeter, and then through all the central roadways, yields plenty of pines and maples, but no oaks. There are drifts of forget-me-nots in the ditches, and the train whistle makes its looping call from the nearby station.

I stop by my brother's grave for a while. My mother is with me and we do a bit of gardening there. She goes off to the water tap to wash her hands and I wander over to a toppled serviceberry tree, dropped by the wind in all its flowering splendour. I mean to take a cutting from it to bring home with me. As I'm walking towards it, I notice how dirty my hands are from the gardening and think that I should have gone with my mother to the tap. But just as I think this, I come upon a stone basin beside one of the graves near the serviceberry that is full of water and I rinse my hands there. The basin makes no sense amid the monuments. It is attached to a tombstone and might be a sort of bird bath, I suppose.

I cut some branches off the serviceberry with my
penknife. Once home and in water their scent will
linger a little longer, and the tree will not be completely
over if some of it still resides with me.

My mother climbs up the stone slabs of the rockery
where the cremated dead are interred, brandishing
an oak leaf. It's brown and cracked, from last fall. She
found it by the water tap, and so we go back there and
follow a trail of old leaves from the pond to the base of
two enormous red oaks, raised up on a little hill. They
are planted close together and are roughly the same
age — not 160 years old, but probably close to a hundred.
Planted perhaps to replace others that occupied the
same spot? We root around at the base of the trees,
pawing through rotting acorns and brittle leaves, but
there are no clues as to the history of the oaks. Still,
nice to stand up with them on their little hillock and
look out over the tombstones and the small ravine
where a trickle of water is crawling down towards
the pond. Nice to listen to the birdsong and watch the
robins perch on the tops of some of the gravestones.
From the oak hillock the fallen serviceberry is a
scattering of white between the dark stones, like snow.

〈〈〈◦〉〉〉

Herbarium Collegii Reginae

Arenaria serpyllifolia L.

Frontenac Co.

Accidentals

specimen

THYME-LEAF SANDWORT
(*Arenaria serpyllifolia*)
Fowler Herbarium

The way through the herbarium is predetermined by the base categories of the plants. It follows a chronology and there are many specimens and collectors in the larger categories, like Ferns or Grasses. But after Oaks there is a winnowing. The broad divisions have given way to more specific ones — from Oaks to individual flowers — and it is hard to determine how to organize my thoughts and impressions as I proceed through the specimens.

But with so much of this project, there seems to be a natural serendipity to things, and if I am just receptive enough to what is actually here, a way through always presents itself.

I have noted down labels and plants that I find interesting in this next section, and when I look at what I have recorded, I realize that there is an order there after all. It is not an order of genus, or even of collectors, but is instead an order that asserts itself from the material I have gathered that has interested me. It is an order of the accidental.

I am led through the half-dozen specimens of wild ginger (*Asarum*), each one, even the specimen from 1880, still smelling pungently of the spice. From there I find several specimens of cannabis (*Cannabis sativa*), including one from Kansas with a label that puts forth a racist assumption, stating that the plant was there in 1968 because of the Mexicans who were employed to build the railroads in the U.S. Perhaps, but this was 1968, the height of the counter-culture movement, when a lot of young Americans were using drugs, so it seems that there are perhaps many explanations for how the *Cannabis sativa* came to be beside the train tracks.

There is a beautiful specimen of wild hops (*Flemingia strobilifera*) from Australia in 1868, collected by an A. Drummond. The specimen loops around the neat ink cursive in the centre of the sheet of paper, which notes the hops to be "Quite Wild," underlined for emphasis. Hops were not native to Australia and would have been brought over on the convict ships in the late eighteenth and early nineteenth centuries. The first cultivated hops were grown there by a James Squire in 1806. He was sent to Australia in the First Fleet for highway robbery and for stealing chickens in London. He grew hops and started the first brewery in Australia, later opening a popular tavern called the Malting Shovel, and finally ending his career as a police constable.[5] The wild hops discovered and collected by Drummond were found on the banks of the Salmon River in the Township of Melbourne.

103

〈〈〈◦〉〉〉

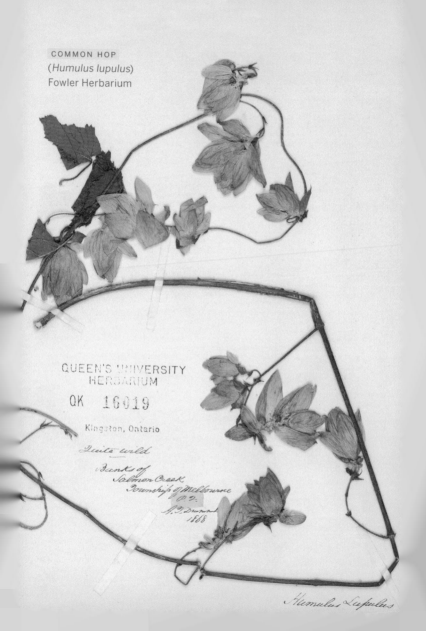

COMMON HOP
(*Humulus lupulus*)
Fowler Herbarium

QUEEN'S UNIVERSITY
HERBARIUM

QK 16019

Kingston, Ontario

Quite wild

Banks of
Salmon Creek
Township of Melbourne
P.2.
J.D. Drummond
1868

Humulus Lupulus

The labels are revealing of many things — the location details of the plant, the priorities of the collector, the story of a particular moment — but sometimes there is an accidental wording that lifts the label more towards poetry. Such is the case of a label for false nettle (*Boehmeria cylindrica*) collected by the Reverend Kenneth James Crawford on December 12, 1970, on the edge of Mill Creek a half mile from Odessa, Ontario. The label is even typed out like a poem:

> *In dense shade at rill-*
> *edge amongst ferns, mea-*
> *dow-rue, baneberry; un-*
> *der white cedar amongst*
> *broken limestone rocks.*
>
> *False nettle.*

Next are two labels for two different plants — devil's thorn and wormseed — one an invasive weed and one a perennial herb, but both brought to North America in the ballast of sailing ships. The label for devil's thorn (*Emex spinosa*) notes that the plant was collected from a "Waste Ground in Pensacola, Florida," on August 8, 1901, and that "The seeds were evidently brought in ballast."

105

Before water was used as ballast for ships, mud, rocks, and earth were shovelled into the hulls from around the harbour where a ship was launched. Seeds and plants were often included in the mix, and when dumped out upon arrival in a different country or continent, these plants took root and thrived. Many of the invasive species in North America, including devil's thorn, were introduced here as "ballast waifs." Now many of these ballast grounds, including the one in Pensacola, Florida, have become archeological sites.

The specimen of wormseed (*Chenopodium ambrosioides*) that is in the Fowler Herbarium was collected in the ballast grounds of the Philadelphia Naval Shipyard on October 19, 1867. This was one of the oldest shipbuilding yards in America. On the shores of the Delaware River, it was the official site for the U.S. Navy and was a very active shipyard throughout the nineteenth century.

FALSE NETTLE
(*Boehmeria cylindrica*)
Henry David Thoreau

My last look at "accidentals" in this section also
concerns two labels, but this time of the same plant
and from the same place. The plant is poverty weed
(*Iva axillaris*) and it was collected from the reservation
of the Pima people on the banks of the Gila River in
southern Arizona. The first specimen of the plant was
collected by George R. Vasey in 1880, and the second
by A.E. Garwood in 1973. Garwood was an employee
of the Fowler Herbarium. The wording on both labels
is similar. Vasey wrote the weed was "Used for food
by the Indians," and Garwood noted that the plant was
"Used by Indians for greens."

107

The Pima, or "river," people, historically lived
around the Gila River and Lower Salt River and sub-
sisted on farming, hunting, and gathering. After the
American Civil War, when their land was in demand
by settlers, they were forced onto reservations in
that area.

What stands out about the poverty weed labels is
that the Pima's use for the plant is prioritized. Perhaps
this is because the weed was collected on the Gila River
reservation and so how the Pima people used it would
be front of mind to the collectors, or they might have
found the fact of the usage interesting. But the labels

were written a hundred years apart and basically say exactly the same thing. The truth is that most, if not all, of the botanists collecting for the Fowler Herbarium were white. So to acknowledge Indigenous Peoples on a collecting label is important and unusual.

Herbaria are now starting to acknowledge and involve Indigenous Peoples. The Kainai First Nation has worked with the herbarium at the University of Calgary to develop an online resource that shares ethnobotanical knowledge about the specimens in the collection. More of this outreach to and participation of Indigenous communities is happening throughout North America. The Navajo Nation Herbarium is a tribal assemblage that is being developed on the Colorado Plateau region. They have collected over ten thousand specimens from the Navajo Nation lands and have a detailed digital database available online. Both the Karuk and Yurok Tribes in California are also building their own herbariums. Their collecting labels include their traditional name for a particular plant, as well as its Latin name, and cultural references and links to oral histories.

Nora Talkington, the botanist who manages the Navajo Nation Herbarium, says of their ever-growing digital database: "Databasing our herbarium has brought our collections and knowledge of the regional flora out of obscurity. We can now easily access all the label information from our collections. This is especially useful for status and distribution reports on rare plant occurrences on Navajo territory, data requests for species lists for specific geographic locations, as well as occurrence oversight."[6]

Herbaria are valuable assets for the sovereignty of Indigenous traditional knowledge.

‹ ‹ ‹ ∅ › › ›

One of the aspects of a herbarium is the condition of the plants when they were collected and preserved. Some plants were picked in their prime, but others were not. Also, some years produced better specimens than others, due to climatic conditions, or the life cycle of the plant itself. So, there is a lot of variation among the specimens.

The collector of seabeach amaranth (*Amaranthus pumilus*) in Atlantic City, New Jersey, writes on the undated collecting label that he perhaps collected his specimen too early.

Fowler mounted an example of thyme-leaf sand-wort (*Arenaria serpyllifolia*) that he found near Clifton, New York, in May 1831 and wrote on the label that perhaps the early season that year had contributed to the fact that the petals weren't as long as the stamen and the leaves on the specimen were imperfect.

Three different labels by three different collectors for the same plant — water chickweed (*Montia lamprosperma*) — take three different approaches in their description of the plant. One notes that it is found "Along brooks and streams" and notes its edible use "in salads." One collector writes a location description that sounds as if it comes from another world entirely. He notes that the plant was found "In frostcrack, next to a large erratic bird rock on open moss-flat tundra" and asserts that it was the only specimen of water chickweed found in that area. But another collector, in exactly the same area, ten years earlier, had found the plant in numerous places "among tall sedges and grasses in damp ground."

〈 〈 〈 ◊ 〉 〉 〉

WATER AWLWORT
(*Subularia aquatica*)
Fowler Herbarium

Bayern.

Subularia aquatica L.

The last "accidental" refers to my accidental discovery from the wording on a collecting label for Juba's bush (*Iresine paniculata*), found in the Santa Catalina Mountains in Arizona in May 1881. The label identified the collector as "J.G. Lemmon and Wife," from Oakland, California.

John Gill Lemmon was a botanist, Civil War veteran, and descendant of the explorer Henry Hudson. He and his wife, Sara Plummer Lemmon — an artist, botanist, and intellectual — spent eight months of every year collecting plants out in the field, only home in Oakland, California, during the winters. They rented their house out when they weren't in it and then lived off that rental income for the winter months.

Lemmon had enlisted in the Union Army after attending university in Michigan. He was captured and incarcerated in the Andersonville Prison, a prisoner-of-war camp so brutal that almost a third of the forty-five thousand prisoners died from disease caused by unsanitary conditions and lack of food, or violence at the hands of the guards or prisoners who had organized into gangs to take what meagre food they could from their fellow prisoners.

When he was released in 1865, John Lemmon went to California to recover his health. While there he met

Sara Plummer, who had travelled from New York City, where she had trained as a teacher and studied art at the Cooper Union. She had worked as a nurse during the Civil War and contracted a bad case of pneumonia; she had moved to California also for her health. She opened a circulating library in Santa Barbara and began botanizing in the countryside and making watercolour paintings of her specimens.

John and Sara married in 1881, and the specimen of Juba's bush in the herbarium was collected on their honeymoon. They were equal partners in their botanical collecting and in the publishing of papers, but the credits always identified them as "J.G. Lemmon and Wife."

Sara Lemmon was, in fact, the more successful of the pair. She was an accomplished painter, and through her efforts, the golden poppy was chosen as the state flower for California. He suffered chronic health problems all of their life together because of his time in Andersonville Prison and died early as a result. Sara outlived him by almost twenty years.

In 1894, an article about the Lemmons appeared in the *San Francisco Chronicle*, under the headline, "Botany is Their Hobby."[7] It described one of their collecting excursions:

Each carried a short Alpine staff, and one of them carries, strapped on the back, a large tin box for specimens. It is also necessary to carry a notebook, and sometimes a blanket, if they intend to remain away over night, for it grows cold rapidly on those great heights after sunset. On these occasions the food for the day is carried in a large pot with a handle and a leathern strap. In this can is a gallon of rather thick cocoa, with crackers stirred in it, which answers for food and drink during the day. Water is out of the question.

〈 〈 〈 ◦ 〉 〉 〉

In the earlier sections of the Fowler Herbarium there were vast tracts of specimens belonging to the same category. (I think of the weeks and weeks it took to get through Grasses, or the long meander I made through Ferns.) But now the categories are changing quickly and I am moving past specimens before I've had a chance to really think about them. One moment there are pond lilies before me, and the next I'm in the midst of delphiniums. So the way I am proceeding has to change as well. It is no longer possible to write about a large group as I am working my way through that group, as I have done until now. The groups have ceded to individuals and the problem becomes how to write meaningfully about the individuals without this book expanding beyond reason and losing its coherence. So, I have had to think differently. I have had to think about what a collection of individuals might have in common, and to look for commonalities in the collectors' labels as well.

What follows are some "escapees," plants that have wandered from cultivated gardens into the wild, or gardens that have become untended and turned wild, or indeed anything that has pushed beyond the boundaries of what had previously contained it.

115

《 《 ◦ 》 》

There is bouncing bet (*Saponaria officinalis*) that has "Escaped from old gardens" in Renfrew County, Ontario, in 1906, and common columbine (*Aquilegia vulgaris*), which is very good at escaping as it was found in a wasteland in the Kingston region in 1967 and was also discovered in "A patch on woods road along Ottawa River" in 1960. Both times it was listed as an "escape" by the collector.

《 《 ◦ 》 》

In Scotland, at Cleish Castle, at an unnamed time and by an unidentified person, winter aconite (*Helleborus hyemalis*) persisted in the site of a ruined garden. Cleish Castle is a sixteenth-century tower house located near Kinross. Its gardens were once extensive and impressive and were restored during the twentieth century to much of their original splendour. I don't know if the winter aconite whose specimen escaped across the ocean to the Fowler Herbarium still grows there, though I am tempted to write to the new owners and ask.

‹ ‹ ‹ ◊ › › ›

In June 1895, some rocket larkspur (*Delphinium consolida*) "Escaped from cultivation" in Knoxville, Tennessee, but somehow retains the brilliant blue of its leaves when pressed. Blue is the colour least found in nature, but it also seems to be the colour that remains most itself after death. Pink and yellow often turn brown, but blue keeps to its original hue.

‹ ‹ ‹ ◊ › › ›

In July 1897, in the fields of Kingston, Surrey, England, some *Dianthus filiformis* was found and was noted as being "Very uncommon there." This flower had been named in 1753 by Carl Linnaeus, but in 1831 it had been renamed as the tunic flower (*Petrorhagia saxifraga*). So the flower had also escaped from its classification, although the British botanist Charles Babington, who catalogued the specimen in the Fowler collection, continued to use the old name. Babington, among other accomplishments, once battled Darwin over a beetle collection.

117

〈〈〈◊〉〉〉

Red baneberry (*Actaea rubra*), a poisonous flowering plant, seems to somehow have escaped from death itself, as the specimen that has been pressed onto its mounting board still has green leaves and fleshy stalks, even though it was preserved in 1948. Only the flattened, dried berries suggest its real age.

〈〈〈◊〉〉〉

WILD LUPINE
(*Lupinus perennis*)
Emily Dickinson

Lupinus, perennis. 16./0.

Greater celandine (*Chelidonium majus*) had escaped from neglected gardens in Kingston in 1986 and was found in open areas near the waterfront.

‹ ‹ ‹ ○ › › ›

Shepherd's purse (*Capsella bursa-pastoris*) was noted on the 1855 label as having been "introduced early into Canada" and was now being found growing wild on July 22 in gardens near Quebec Street, also in Kingston.

‹ ‹ ‹ ○ › › ›

Yellow alyssum (*Alyssum calycinum*) was discovered on a Michigan College campus by W.J. Beal on June 20, 1893, who noted on the label that he thought it had got there by being "introduced in lawn-grass seed."

William James Beal was a Harvard student while Thoreau was still alive and three years after Charles Darwin had published *On the Origin of Species*. He is best known for hybridizing corn. He also created a botanical garden on the grounds of the University of Michigan campus that is still there. He began an ongoing experiment in 1879: he buried vials of twenty-one different kinds of seeds that are periodically dug up and planted, to see how many will grow. The experiment is destined to run until 2100.

‹ ‹ ‹ ● › › ›

In rocky outcrops and crevices north of the Queen's University biological station, the garden escapes: goldmoss stonecrop (*Sedum acre*) and orange stonecrop (*Sedum kamtschaticum*) were observed growing in 1960 and 1968, and a white currant (*Ribes sativum*) bush was found there growing under a red oak in 1964. Meadowsweet (*Filipendula ulmaria*) also persisted in 1968.

Queen's University had Canada's first botanical garden spread out over a seven-acre site on campus. The makeup of the garden, in terms of plants and seeds, was provided by botanists, such as Asa Gray, and by the citizens of Kingston. George Lawson, one of the Fowler Herbarium's early collectors and the founder of the biology department at Queen's University, used the plants in his teaching. The garden was home to many plants with medicinal uses, including perhaps the sample of lady's mantle (*Alchemilla vulgaris*)

found on the university lawn on August 25, 1964, by
Harold Zavitz, who notes on the label that the plant
was "probably a relic of the Old Botanical Garden."

Lady's mantle has historically been used for regulating menstruation, easing childbirth, and treating
wounds. It was also used for softening dry skin, healing
mouth ulcers and bleeding gums, and treating diarrhea.

〈 〈 〈 ◦ 〉 〉 〉

A bird of paradise (*Caesalpinia gilliesii*) bush, native
to South America, was found in Tucson, Arizona,
in 1972 and was branded as an escapee, but noted by
the collector, A.E. Garwood, that it was "Becoming
naturalized in many places."

〈 〈 〈 ◦ 〉 〉 〉

White clover (*Leguminosae melilotus*) had escaped
from a garden in Renfrew County in June 1906, and
crownvetch (*Coronilla varia*) was on the run in Ottawa
in June of 1952.

〈 〈 〈 ◦ 〉 〉 〉

121

Botanical Name. Rosa rubiginosa

«‹‹‹‹‹ • ››››››

The
Rose Family

SWEETBRIER ROSE
(*Rosa rubiginosa*)
Fowler Herbarium

I am now moving through the Rosaceae category and am looking forward to viewing the roses. But there are a lot of plants in this category before I get to actual roses. There are stone fruits, like chokecherry, and crabapple varieties. (W.G. Dore tasted the sample of a paradise apple [*Pyrus malus*] he found near Prescott, Ontario, in October 1960 and pronounced the fruit "hard and sourish.")

A rose is not a rose at all, it seems. It is a cotoneaster, and a hawthorn, and a serviceberry, and a potentilla. It is also a bramble. Or not. The Reverend Kenneth Crawford described the northern dewberry (*Rubus flagellaris* Willd.) in June 1969 as "Very difficult to identify: specimen incomplete: could be some other bramble."

When I do finally encounter some roses, they are brief as actual, living roses, flashing past me in their dried state, their leaves salmon pink and crisp as crepe paper.

‹ ‹ ‹ ◊ › › ›

WILD BLACKBERRY
(*Rubus occidentalis*)
Henry David Thoreau

I am always grateful to encounter a label from W.G. Dore because, now having looked through almost half of the specimens at the herbarium, his descriptions remain the most detailed and vivid. Also, he often includes photographs, which I find fascinating. For the specimen of shrubby cinquefoil (*Potentilla fruticosa*) he collected along the Ottawa shoreline in 1966, he included a black-and-white snapshot of the bush growing on a limestone slab. The small flowers look like miniature stars against the dark mass of the shrub. In the upper right corner of the photo is a slash of overexposure from the bright sun on that August day.

It is with joy, and then dismay, that I find an actual letter from Dore, tucked into a specimen file. The typewritten letter is addressed to his fellow collector at the herbarium, Gar Garwood, and the letter is dated July 28, 1989. Dore would have been seventy-seven at this point — he died seven years later in 1996 — and he apologizes in the letter for his failing memory.

WILD PLUM
(*Prunus americana*)
Henry David Thoreau

"Memory is getting loose," he writes, in his exacting way. The subjects in the letter are a trickle-down of his thoughts. He asks Garwood to correct a mislabelling on a specimen of ivy. Then he talks about Beschel and his research on the lichens of the Viking tombstones in Greenland. Next he laments that the old botanical garden at Queen's has been turned into a student parking lot. He talks about elm disease, and the first microscopes at Queen's, brought there by George Lawson from Scotland. Then he pinpoints a spot on Montreal Street where native plants were found in 1834, before moving swiftly on to reminiscing about old growth forest being cut down around the same time. He ends the letter with an observation that the mouth of the Cataraqui River is filled with what looks like oil or sludge but is, in fact, bladderwort. All this string of unrelated information is on a single typewritten page.

Earlier in the day, I had read his label for Carolina whitlow grass (*Draba reptans*) — despite its name, not a grass at all, but a wildflower from the mustard family — from June 15, 1963, with my usual pleasure. "A springtime ephemeral on limestone pavement," he wrote, "probably abundant, but very difficult to see."

127

〈〈〈◇〉〉〉

part three

Summer

Chicory

CHICORY
(*Cichorium intybus*)
Helen Humphreys

At exactly this time in my research and writing, my dog, Charlotte, died of a swift and sudden cancer in her heart. She was only nine years old and was my constant companion. Her loss not only could not be ignored, but upended everything in my life, including this book.

She was diagnosed on a Sunday in July and euthanized at home three days later. Some of that time was spent with x-rays and ultrasounds, and in giving her a small operation to drain the fluid around her heart. This operation gave her a full day of being entirely herself again, before the sac around her heart filled up once more and I had to have her euthanized, rather than risk the possibility of her bleeding to death internally and dying in agony.

On that last day, we did what we usually do; we went for a walk with our human friend on the piece of countryside where we always walk, every day, in all seasons, without fail. It is the personal paradise I wrote about at the beginning of this book.

On this day, a Tuesday, we arrived at the conservation area at around seven thirty in the morning. It was sunny out, but not humid, a perfect temperature for walking. We piled out of the car and immediately Charlotte started hunting in the tall grass that bordered the edge of the first field. She has always loved hunting, and hunting voles is her favourite thing on Earth, because voles are plentiful and easy to catch. I think, from watching how she hunts them, that I could probably catch one myself, if I wanted to. They seem numerous and extremely slow-moving.

The dog's hunting this morning was very deliberate and purposeful. She was intent on what she was doing, to the exclusion of all else, and we could tell, from this change in her manner, that she was focusing very hard on having this time of concentrated pleasure, of being fully herself.

On the path, at the start of our walk, there were a pair of cardinals, then a pair of downy woodpeckers, also many yellow warblers. The air was churning with coloured birds and the wheel of their songs.

Near the place we call the wallow — because it is a deep hole in the marsh grasses that is often filled with water, through which the dog joyfully cavorts — we met some of Charlotte's human friends, three women whom we see almost every day and who love the dog. They cried when we told them that this was probably her last walk, and they thanked her for the way she always greeted them, for her cheerful and ebullient spirit.

FIELD THISTLE (*Cirsium discolor*) Helen Humphreys

Further on there was a rabbit to hunt. Charlotte didn't have the energy to do a full-on chase, but she pointed it for a long time, and then gave a small charge to force it to leap into the bushes for safety.

When we hit Carnage Alley, it was obvious that the walk was special. Made more so, of course, because it would be our last walk together, but also because everything had conspired to make it so joyful and beautiful. The milkweed had just come into flower in the field and their scent filled the air we walked through with a strong, heady, sweet musk. There were bobolinks swaying on the old milkweed stalks, and eastern kingbirds overhead. Along the edges of the field was a blue wash of chicory.

My human friend had tried to tempt Charlotte to eat by offering up her toast with peanut butter, but the dog was refusing food at this point and had turned her nose up at it in the car. But here, at the beginning of walking through the field, she took the toast as a prize, carrying it all the way back to the car, as she would have carried an animal that she had hunted and killed. It seemed like another way she was trying to remain herself, to enjoy herself as hard as she could on this last day.

We passed some people on our way across the field. They laughed at the sight of the dog holding her head high, carrying her piece of toast, and that laughter was as beautiful as the birdsong. Sometimes Charlotte stopped and turned on the path to make sure we were still behind her, and all of her body language, when she turned and then continued on, was jaunty and joyful. There was no way to not be happy on this walk.

When we negotiated the final corner into the last patch of field, there were three young stags grazing in the tall grasses to our right. They were about twenty feet away from us, unconcerned by our presence and by the presence of the dog, who noticed them but chose to ignore them. She could feel good about transporting her piece of toast to the car, but she didn't have the energy to chase the deer, and she was all about feeling good this morning, not about reminding herself of what she could no longer do.

QUEEN ANNE'S LACE
(*Daucus carota*)
Helen Humphreys

We had never seen three stags together before. Ever. Not in all the years we had been coming to this place, and it was hard not to connect it to the dog, and to the glory of her departing spirit. It was hard not to think that the stags were there to announce Charlotte leaving this Earth, that they were part of this strange and beautiful, almost otherworldly, procession through the July morning.

When the dog was very young, she and I had been walking along a forest path and as we turned a corner on the path we had seen a young stag up ahead of us. The dog was too much of a puppy still to have wanted to hunt it, or chase it, and we all walked peacefully along the path together — the stag ahead of us, and then the dog with her loose-limbed youthful gait, and then myself. So, the three stags at the end of Charlotte's life made sense to me in terms of this visitation at the start of her life.

A stag in many traditions is the link between this world and the next, is the symbol of the "thin place," where the membrane of reality is at its most porous.

The dog of a writer, as a friend of mine pointed out after Charlotte had died, is different from other dogs, or the relationship is different, more intimate, because the dog is allowed into the private space of writing that no person is ever allowed into. In that way, Charlotte spent more hours with me than any other creature or human. She was there, curled up on her end of the couch, snoring through my morning's writing, pretty much every single day of those nine years together.

‹‹‹•›››

But it is not our history together that has changed my course in this book. It is not even missing Charlotte and having those feelings flood over everything else and render it essentially meaningless. No, it is that last walk of hers, that beautiful parade through the field, with the birds and the stags and the scent of the milkweed, and Charlotte leading us along with that piece of toast in her mouth. It was how she knew that she was dying and how hard she tried to live in those moments she had left. It was the heroism of that act, of celebrating life so deliberately overtop of a death that was merely hours away. In the car ride home, I opened the window for her and she thrust her head out (something she usually did not do) and breathed in great lungfuls of the sweet morning air.

So, after witnessing that, after being a part of it, how can I turn away from the sunshine and the flowers and the wheel of birdsong towards the cold cabinets filled with folders of dried and dead plants?

I can't.

‹ ‹ ‹ ◦ › › ›

For the past couple of years I have been teaching myself how to draw, bringing home wildflowers or twigs and seeds from the forest floor from my morning walk, and then trying to render them — not as ideal examples of a particular plant, but just as I have found them — an acorn cap, a broken pine cone, a lichen-covered stick. I like taking the long look at the small thing. No piece of nature is insignificant and all of it is worth memorializing.

Which, I suppose, is what I realize with my dog's death. This world I live in — the world that you live in — is a world of disappearing species, but it is also still a world of wonder and beauty. And while we all must do more, and petition our governments to do more about the climate crisis, and not ignore the fact that humans are responsible for the destruction of species and habitat, we must also celebrate what is still here with a ferocious reverence.

Charlotte died at the time of chicory, milkweed, and fireflies in southern Ontario, and also the time at which Queen Anne's lace was just coming into its own. At first, I try to draw chicory (*Cichorium intybus*), for days at a time. It is very hard to match the exact blue colour of the flower. As I mentioned, blue is the hue least seen in nature, and this particular blue has underpinnings of both indigo and lavender. It also fades over the course of a day, and so it is hard to know which is actually the true colour: is it the darker blue of the morning, or the faded lavender in the middle of the day, or the almost white at the end of the day? The colour drains from the flower all day long, and the several stalks that I picked to put in water and draw keep changing by the hour and make it impossible to pin down. The flower stalks at the edge of the field have many flower heads on one stalk, and each flower head opens for only part of one day and is then finished, with the whole stalk producing many flower heads over the course of the summer. It is a bit like the way the fireflies spark across the dark lake of the night field, one coming on while one is going out, and I wonder at the relationship between these things. What will I find if I look for connections between the four things that were in the ascendant when Charlotte died — chicory, milkweed, Queen Anne's lace, and fireflies?

139

〈 〈 〈 ◇ 〉 〉

CHICORY
(*Cichorium intybus*)
Henry David Thoreau

Chicory, a member of the dandelion family, is native to Eurasia and has traditionally been used as a coffee substitute, a salad green, and as medicine to cure jaundice, gout, rheumatism, high cholesterol, rapid heart rate, and in poultice form to relieve swollen eyes. It is also thought to be a good treatment for liver enlargement. During the American Civil War, the roots were roasted and ground to produce the coffee substitute that was very popular among the soldiers, when coffee itself was in short supply.

In 1748, Carl Linnaeus developed a "floral clock" in Uppsala, Sweden, where he noted the different moments of the day when various flowers opened, the idea being that one could tell the time by which flower was open. Linnaeus placed chicory at the 5 a.m. position on his theoretical clock. In Uppsala, the flowers closed at 10 a.m., but here in North America, they remain open into the afternoon.

The Greeks and the pre-Celts both envisioned chicory as a devoted maiden, waiting for her lover to return. The blue of the flowers was meant to be the colour of the eyes of the maiden, and the flower has come to symbolize faithful love, or waiting in vain.

‹ ‹ ‹ ◊ › › ›

After failing all day to draw the chicory with any accuracy, I drive out to the field after dark to look at the flowers and see what state they are in at night.

There are fireflies along the edge of the road and in the ditches where other swathes of chicory grow, but the field itself is not lit by fireflies. It is interesting that both the fireflies and the chicory keep to the edge of the field, and are filling the ditches on the road in. Maybe they do have something in common besides the briefness of their brightness?

I stop the car and turn off the engine and get out into the mosquito-filled night. With a flashlight, I move along the edge of the field and look at the chicory stalks. Each flower head is closed into a tight knot and those that have bloomed already are barely distinguishable from the new buds, but looking closely, I can see that the knotted old blooms are withered and white. They will drop off the stalk during the night, or in the morning, the shortness of their lives so shocking, a crazy, sped-up version of our own perhaps.

Now that I have witnessed the chicory at night, with the blooms at the end of their lifespan, it only makes sense to go to see them at dawn, when they are just beginning.

The flower heads unknot imperceptively, their deep

blue colour at sunrise a reminder of the night that has just passed, for the blue is reminiscent of the shade of "the blue hour" or the last vestiges of twilight in the evening. The slow uncupping of the flower heads happens at different rates on a single stalk, with the flower heads that are further up the stalk, and therefore closer to the sun, opening at a faster rate than those flower heads near the base of the stalk. The flower heads all turn to face the sun as they open, and part of the effect of a stand of chicory is the wash of blue that is a result of all the open flower heads facing in the same direction. It seems to me that telling the time by chicory is a less accurate notion than being able to determine the compass points by the inclination of the flower heads.

When I go back to my desk and attempt to draw it again, I layer the deep blue of the morning into the centre of the flower heads, fading the colour out into the ends of the petals, trying to capture the nature of the plant and the movement of colour. It's closer than my original attempt, but maybe the colour is moving with such quickness that a true likeness is impossible and there can only ever be an approximation? Maybe chicory is one of the few places where colour can be seen filling and draining, moving like an arpeggio up and down a single stalk of flowers.

‹ ‹ ‹ ◊ › › ›

During the winter and spring, I had been making notes about milkweed (*Asclepias*) because I was interested in the way the stalks persisted in their upright position all through the colder months. I was interested in the nature of collapse, and how long something could hold its form. In the winter field, milkweed was the only vertical, and even though it was exposed fully to the winter winds and the scouring needles of snow, it remained upright for the whole of the season. Collapse, in its inverse, is actually the study of resilience.

I wondered too about the relationship between the harrier and the milkweed. Is it accidental that the empty milkweed pod, after the seeds have been released, resembles a mouse or vole? The empty pods would be very visible to the harrier, or to a hawk that was hunting above the field in winter. Does the predator use the pods to find its prey?

Mice eat the monarch butterflies that visit the milkweed plants in summer. They also feed on the chrysalis of the monarch, and they use the seeds and floss of the milkweed to line their nests, so it follows that hawks and harriers could use the shape of the milkweed husks to find the rodents. The fact that the mice use the plant so liberally suggests that they also remain in the vicinity of the milkweed, perhaps burrowing at the base of the plants.

Humans have made much use of the milkweed plant through time, mixing the silken fibres with wool or flax and weaving this into cloth, and stuffing pillows and mattresses with it as well. Paper can also be formed from the tough, fibrous stalks, and during the Second World War, the hollow, silky fibres were coated with wax to make them buoyant and waterproof and then used to stuff life jackets for the Allies.

In July, this July, the dusky pink globe-shaped flower heads are heavy and fragrant, drooping on their stalks and surrounded by the fluttering of monarchs, who seem, thankfully, more plentiful in the field this year than last.

In the record that I kept of the progress of the collapse of the milkweed stalk, the plant remained upright into April, finally succumbing to rot from the months of snow and the relentless spring rain. But by the time the stalks collapsed, there was already new growth showing at the base of the plant, and before they collapsed, the stalks were the spot where the returning bobolinks (whose numbers are much depleted) alighted after their astonishing twenty-thousand-kilometre round-trip migration — one of the longest migrations of any songbird.

As soon as the July flower heads droop and then shrivel up, the bobolinks again use the stalk to perch on, as it is the tallest plant in both the summer and winter field. The milkweed, perhaps more than most flowers, is the centre of a community that includes birds, insects, and mammals, and is its own small ecosystem.

〈〈〈๑〉〉〉

In my study of the field, I searched for and read many different field guides to North American wildflowers. Each one differs from the next in subtle and not so subtle ways, and the guide I found myself gravitating to the most was one of the earliest ones — *How to Know the Wildflowers* by Mrs. William Starr Dana, published in 1893 and selling out its first printing in five days.

Mrs. Starr Dana was the pen name of Frances Theodora Parsons, born in New York in 1861 and privately educated. When she was twenty-three, she married the naval officer William Starr Dana, who died in a flu epidemic six years later. To console herself through her grief, Parsons went for long walks in the countryside with her friend Marion Satterlee, who happened to be a botanical illustrator. The idea for the field guide was born during those walks, and the two women worked together, with Parsons writing the text and Satterlee providing the many pen and ink illustrations.

Frances Parsons wrote other books about wildflowers and ferns, and went on to marry again, although her second husband also died after six years of marriage, in an automobile accident. Parsons moved from the country to New York City after his death and was active in politics and women's suffrage. She lived to be ninety years old.

What makes *How to Know the Wildflowers* my favourite of the field guides is its approach to its subject. Firstly, Parsons grouped the guide by colour, which is the way many of the modern wildflower guides are also organized. The flowers then move through each particular colour according to when they come into bloom. The information that Parsons provides about each flower is a combination of fact and whimsy. She sometimes quotes a bit of poetry, or refutes something someone else has said. She argues with Emily Dickinson, who states, "Nature rarer was yellow / Than another hue." Parsons points out beneath this quote, that yellow is quite the commonest colour in nature.[1] She also acknowledges the Indigenous names and uses for various flowers, often placing the Indigenous name for a plant in the header, where she has also put the Latin and common names. This is the only field guide written by a settler where I have found this kind of acknowledgement.

〈〈〈◇〉〉〉

Here is Frances Theodora Parsons's description of Queen Anne's lace (*Daucus carota*):

When the delicate flowers of the wild carrot are still unsoiled by the dust from the highway, and fresh from the early summer rains, they are very beautiful, adding much to the appearance of the roadsides and fields along which they grow so abundantly as to strike despair into the heart of the farmer, for this is, perhaps, the "peskiest" of all the weeds with which he has to contend. As time goes on the blossoms begin to have a careworn look and lose something of the cobwebby aspect which won them the title of Queen Anne's Lace. In late summer the flower-stalks erect themselves, forming a concave cluster which has the appearance of a bird's nest. I have read that a species of bee makes use of this ready-made home, but have never seen any indications of such an occupancy.

This is believed to be the stock from which the garden carrot was raised. The vegetable was well known to the ancients, and we learn from Pliny that the finest specimens were brought to Rome from Candia. When it was first introduced into Great Britain is not known, although the supposition is that it was brought over by the Dutch during the reign of Elizabeth. In the writings of Parkinson we read that the ladies wore carrot-leaves in their hair in place of feathers. One can picture the dejected appearance of a ball-room belle at the close of an entertainment.[2]

149

‹ ‹ ‹ ◊ › › ›

Queen Anne's lace, like milkweed, has a summer and winter persona. In winter, it curls up into a little bird's nest ("bird's nest" is indeed one of its common names), its stalk and fronds dried to a coppery colour, but still the delicate threads of the blossom persisting. It is easier to draw in its winter guise, and it was one of the first flowers I ever drew because the balance between the shape and the small stars of the dried umbels is not difficult to render. But when I try to draw the flower in summer, I find it almost impossible.

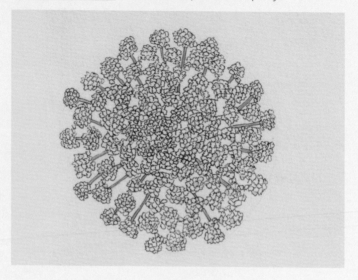

150

I am not interested in botanical illustration as such, in that I don't want to draw something that so closely resembles its model that there is no expression in the drawing. So, what I try to do is to look for the pattern in a particular piece of nature — for there usually is one — and be faithful to the pattern, but also make some interpretations beyond that. For example, I drew a bulrush with its fibres bursting through the casing like an overstuffed sausage, and I drew a chestnut with its spikes in imitation of a medieval weapon.

‹‹‹◊›››

The Queen Anne's lace in the field move almost
constantly, bobbing their heads on their long, slender
stalks. From afar, they are easier to see than other
flowers, because their blaze of whiteness is so
bright. Up close, the flower head is a series of closely
connected umbels, each densely packed with tiny
star-shaped blooms. The blooms and the umbels are
easy enough to draw, but what proves difficult is
getting the distance between them right. There has
to be space between all of the parts in the head of a
Queen Anne's lace, even though it looks to be of one
piece. Also, every single flower is different. Some
have more space between the umbels, some have less.
Some of the flower heads are large, and others small,
so there is no definitive exactness to the appearance,
but there is also no latitude for getting it wrong. If the
spacing is off by any amount, the whole of the drawing
looks completely inauthentic. Perhaps this is because
the flower head, when viewed close up, is abstract, and
in making something abstract, there is little room for
anything that does not feel true.

What I have learned from drawing flowers is that
each one, while subscribing to a pattern, is also an
individual, and when I am drawing it, I have to pay

151

attention to the character of each flower and present that in the finished piece. And, just as I am drawn more to the character of some people, I also prefer the character of particular flowers, and in Queen Anne's lace, I prefer there to be space between the blooms and the umbels, for the head of the flower to have an open appearance, the "lace" loose enough to see through to the field grasses below.

Drawing is, of course, mostly looking, or an excuse to look long and hard at something. Francis Hallé, a botanist who also draws, says, "The extended time required for drawing amounts to a dialogue with the plant ... Drawing represents the work of human thought."[3]

‹ ‹ ‹ ◦ › › ›

The act of drawing has increased my powers of observation, and while I have been paying attention to individual flowers, and specific flower varieties, I have also started to notice accordances between certain aspects of nature. Chicory and fireflies have something in common, as do monarch butterflies and orange day lilies. The butterfly wing and the bracket mushroom have something in common, as do a wasps' nest and a barred owl, and the rosehip and the octopus, the shape of the butternut pod and the shape of an owl.

While animals and plants are still classified using a version of the Linnaean system[4] I prefer a more interconnected way to look at plants and animals, and in my own practice of thinking about, and drawing, the natural world, I am trying to make connections — not so much a way of delineating a direct line between one thing and another, something more subtle than that: colours that line up, a way of movement that resembles another way of movement, a shape that crosses from one species to another.

And when I think of field guides, I think of making small guides to both the tangible and the ephemeral, such as a *field guide to memory* that shows the angle of the rush after the blackbird has lifted, that shows the strew of apple blossom after the storm has passed, the closed head of a flower after the rain has ceased.

153

‹ ‹ ‹ ❂ › › ›

SWAMP MILKWEED

(*Asclepias incarnata*)

Henry David Thoreau

It is the middle of August now. The plump globes of
the milkweed flowers have drooped and dried up.
The chicory is flickering out at the edges of the field and
in the ditches. The fireflies have gone. Queen Anne's
lace still nod their heads in the late summer sun, but
there is now an emptying out of the field in preparation
for fall.

My dog has been dead for over a month, and while
I have tried to keep this space open for her for as long
as possible, to keep my connection to her memory
strong, and the spirit of her last, triumphant, beautiful
walk through the middle of summer alive as a beacon
to guide me forward, all of this is getting harder as the
plants that kept her company in life are fading from our
shared landscape.

Even the wallow, where there is still an imprint
of one of her paws, has dried out and filled in with the
stiff bristle of field grasses.

The flowers of Charlotte are leaving, and perhaps
that means it's time for me to move indoors, and go
back to the herbarium.

‹ ‹ ‹ ● › › ›

Order *Ericaceae*

Heath Fam

Name *Epigaea repens*

Trailing Arbu

Locality *Wicklow pin*

Date *May 2, 189*

Wicklow Twp

N—

RECORDED IN
NAT'L HERB. CAN.

‹ ‹ ‹ ‹ ‹ • › › › › ›

Trailing Plants

specimen

TRAILING ARBUTUS
(*Epigaea repens*)
Fowler Herbarium

I am into the trailing plants now — vetch and sweet pea — following a series of specimens and collectors and finding much that is elusive.

My research into the different collectors depends on there being a trail — an article written about them, an obituary, memories from colleagues and students, authorship of a book — but there often isn't something definitive, and so I can't say anything about a person who went out one July morning and picked an example of vetch (*Vicia*) from a field near where they lived. This is particularly true when the name on the collecting label is prefixed by "Mrs." and includes no first name. Who was the Mrs. Russell who took both vetch and sweet pea (*Lathyrus odoratus*) from Lake Saskatoon in the summer of 1919? Who was Mrs. Terry from New Hampshire, who in August 1902 decided to preserve a specimen of yellow wood sorrel (*Oxalis stricta*), and how did it end up in the collection of this herbarium? More puzzling is the beautiful, laboured, calligraphic script of a Mrs. Godd, who in July of 1868 captured a specimen of purging flax (*Linum catharticum*) in Kirkton, Scotland, noting on her label that it had been found in "Dry pastures."

The mysteries are as many as the tangible results, and they are often attached to the women collectors, about whom little is found because women are mostly not included in the historical record. Even more elusive than a label with "Mrs." on it is one with "Miss." Were these mysterious Misses young women or older un-married women? Were they students or self-taught? Did they still live with their parents, or were they self-sufficient?

In the historical record there is no equality between men and women, but in the herbarium, everyone is equal and there is no difference between the attention given to Mrs. Godd's specimen of purging flax and that of the male botanist, M.L. Fernald, which follows hers in the file folder.[5] I appreciate this democracy of categorization, and I have come to think of the herbarium not so much as a wilderness but as a garden of sorts, with equal reverence given to all the plants, and to all the collectors, with no one and nothing preferred above anyone or anything else.

YELLOW WOOD SORREL
(*Oxalis stricta*)
Emily Dickinson

There is no one whose story I can unfold in Vetch and Sweet Pea, but in Panicled Tick Trefoil (*Desmodium paniculatum*) there is Jack Gillett, who was a botanist working in Ottawa through much of the twentieth century, an expert on gentians and clover, who made many field surveys with W.G. Dore and was described in his memoriam as having loved playing the piano, studying languages, collecting stamps and minerals, and skinny dipping.[6]

For what are we remembered, and what has been forgotten? Jack Gillett loved to picnic in cemeteries and swim naked. Mrs. Godd, whoever she was, might have also enjoyed similar pursuits, but all we know about her for sure is that she was in a dry pasture in Scotland on a July day in 1868, and that she had beautiful handwriting.

《《◇》》

Some people aren't remembered and others choose to memorialize themselves. A woman in Nova Scotia sent in a specimen of bird vetch (*Vicia cracca*) from her own garden in 1983, noting on the label that it was "Mixed in with regular purple-flowered plants in unweeded flower garden on south side of home."

In a New Brunswick garden in 1971 a vine (hog peanut) was accidentally discovered residing in the centre of a Japanese iris and sent in to the herbarium by the woman who owned the garden.

In Lethbridge, Alberta, the story of a specimen of common stork's-bill (*Erodium cicutarium*) was that it was found in a "Gravel bed by parking lot of Lethbridge Tennis Club. Very stunted plant due to constant mowing." The collector was perhaps a member of the tennis club and therefore knew that there was "constant mowing" in the area around the parking lot where the plant lived, and where he passed by it on his way in and out of the club.

The more common the plant, the more common the location where it is found. There is no need to do a field expedition to discover vetch or sweet pea or flax. They are in our gardens and at the edges of parking lots. They trail around fences and up the sides of abandoned buildings. There is not an effort required to find them, although to notice any plant takes a certain amount of curiosity and observational skill. Not everyone coming and going from the tennis club would have been bothered to record the common stork's-bill, or make the note about its stunted condition. The trail through the herbarium is a path along which there have been many good observers and devotees of the plant kingdom.

part four

Autumn

Spurges & Worts

GRAY GOLDENROD
(*Solidago nemoralis*)
Fowler Herbarium

Summer seems to have given way to fall without any struggle. On my drive up to the herbarium this week, the fields are gold with ragweed and goldenrod. The asters are just beginning, the apple trees are filling with fruit, and I see three fawns at the edge of a meadow and some wild turkeys in a ditch.

The plants may switch out in the fields, but inside everything remains as it was. The plants there are static rather than fluid, and yet still seem somehow, oddly, alive. Perhaps because they are still taking up space in the world? In fact, a dead specimen takes up exactly the same space as a live plant does. In the winter field, the dead stalks of milkweed were the only verticals, and, although dead, they seemed alive because they occupied space in the same way they had when they were alive.

A specimen of prickly ash (*Zanthoxylum americanum*) from 1891 is still razor-sharp and I have to be careful not to cut my fingers on it.

‹ ‹ ‹ › › ›

MILKWORT
(*Polygala comosa*)
Fowler Herbarium

Vetch and sweet pea have been left behind, and now I am traipsing through a multitude of spurge (*Euphorbia*) and milkwort (*Polygala*) by way of the creosote bush collected by a Marcus E. Jones in the Mohave Valley, California, on May 3, 1884.

Creosote (*Larrea tridentata*) has long been used by the Indigenous populations of North and South America to treat everything from tuberculosis to arthritis, as well as serving as a hair tonic and antiseptic. More recently it has been used in AIDS and cancer treatments as it traps free radicals and detoxes the blood of the patient.

Marcus E. Jones, who was described in his memoriam as a "curmudgeon,"[1] made his living collecting plant specimens from the western United States and then selling them on to collectors and institutions. He was a loner, having fallen out with many of his fellow botanists, and he spent most of his time outside, wandering through the Utahan and Californian landscapes pushing a handcart loaded with camping, photographic, and botanical supplies. His wife, Anna, whom he rarely saw, and who eventually left him, supported his peripatetic lifestyle by working first as a schoolteacher, and then by running a boarding house and taking in sewing, all while raising their three children.

Jones had his own botanical journal, *Contributions to Western Botany*, and he published a monograph on the genus *Astragalus* (a herb used in Chinese medicine), as well as producing tourist booklets for Salt Lake and Utah. He also promoted mining and was sometimes an expert witness in court cases about the negative effects of smelter smoke.

Eventually trading in his handcart for a Model T Ford, Jones died in an automobile accident in 1934, on one of his endless collecting excursions.

Nowadays, someone like Marcus Jones would not be given authority in the realm of science, but I like how a homeless man with a handcart full of belongings in 1884 was a respected contributor to the botanical world. In fact, it seems that many of the amateur botanists lived a life akin to artists, that there was more room for eccentricity then.

‹ ‹ ‹ ◦ › › ›

By this point, I have roughly looked through well over half of the 144,000 specimens, and I am so used to how each one is generally presented on the page that it is very obvious when something is different. Usually each specimen sheet contains the plant, glued down or affixed with tiny strips of adhesive tape, and the collector's label in the bottom right corner, which is about the size of a business card. But the specimen of drumheads milkwort (*Polygala cruciata*) that I find next has a multitude of labels covering the bottom section of the page, as well as a paper restaurant napkin stuck on top of the specimen with the words "extinct? mortem" inked on it in blue ballpoint pen.

The original label is the standard one for the Fowler

Herbarium and documents that this drumheads milkwort was found by a W. Cody on August 9, 1896, in a "Chestnut Grove" in Windsor, Ontario. On a label to the left of this first label, dated 1963 by a B. Boiven from the Ministry of Agriculture, is a dispute of the milkwort's origin. "Could this have been picked up while the collector was holidaying in the States?" writes Boiven, noting also that the plant "Does not seem to have been seen since." On a label below that one he lists the possible identity of W. Cody as being a high school principal in Windsor and an amateur naturalist. In different handwriting underneath that is written "And later physician at Hamilton, Ontario." On yet another label on the same specimen sheet, from 1984, a botanist from the National Herbarium in Ottawa has noted the varietal of this milkwort as *aquilonia* and included the specimen in the "Rare and Endangered Plants Project."

It's like a novel in précis and I wonder, for the first time, about the accuracy of collectors. While it seems pointless to lie about where a specimen was found, did they always tell the whole truth? Did they sometimes misremember where they had collected a particular flower? Up until this moment I have thought of the label and the plant specimen as being closely connected, as belonging together, but perhaps sometimes they have

CROSS-LEAVED MILKWORT
(*Polygala cruciata*)
Fowler Herbarium

QUEEN'S UNIVERSITY
HERBARIUM

GK 33214

Kingston, Ontario

CAN Druebine
1985

Could this have been picked up
while the collector was
holidaying in the States?
Does not seem
to have been
seen since

Revisé par B. BOIVIN.
Ministère de l'Agriculture, Ottawa.

RARE AND ENDANGERED PLANTS PROJECT

Polygala cruciata L.
var. *aquilonia* Fernald & Schu.

Kathleen M. Pryer
National Herbarium of Canada   1984

FLORA CANADENSIS

EX COLL. J. FOWLER   Essex

var. *aquilonia* Fern. & Schu.

? = W. S. Cody ?
High School Principal
1913 at Windsor

Polygala cruciata L.

no real relationship at all. Some of the information on the collecting label about the plant is subjective and dependent on a variety of factors, including the honesty and sanity of the collector themselves.

I can't find much to add about W. Cody, except that he was also, apparently, interested in butterflies, as he donated a specimen of a drury (*Argynnis idalia*) to the Entomological Society of Ontario in 1896, the same year he collected the much disputed milkwort.[2] Butterflies are attracted to milkwort and perhaps Cody was not after it at all when he picked it, but was rather following the trail of a wandering butterfly?

〈 〈 〈 ◦ 〉 〉 〉

Just when I had almost forgotten him, W.G. Dore shows up with two specimens of the same plant: Virginia threeseed mercury (*Acalypha virginica*), a plant in the spurge family. They seem to be from the same collection excursion as they are both from Prescott, Ontario. One was taken on September 14, 1960, and the other was collected a week later. The note on the first specimen has the rather interestingly worded location "Around dump heap in Blue Church cemetery."

〈 〈 〈 ◦ 〉 〉 〉

The birds have mostly left the autumn woods when I walk in the mornings. There are spiderwebs slung between the milkweed stalks in the meadows, and dew on the grass at the start of the day. I have been watching the emergence of a puffball mushroom (*Scleroderma citrinum*) in the woods at the biological station, admiring the power with which it bursts through the loam and pine needles, the swiftness with which it grows enormous, glowing between the trees like an albino football.

PUFFBALL MUSHROOM   (*Scleroderma citrinum*)

AMERICAN BLADDERNUT
(*Staphylea trifolia*)
Fowler Herbarium

Staphylea trifolia L.

03 08 1969   S. VanderKloet
S. VanderKloet   1969 1
31C08E   065/170
  Ont., Leeds Co., Front of Leeds Tp.
  Abundant on the edge of a prunus (frax
stand.

At the herbarium, I am looking through samples of bladdernut (*Staphylea trifolia*). The Reverend Kenneth Crawford — who lived to be 101 and collected his plant specimen near Odessa, Ontario — wrote on the label that "Fruits persist on trees in winter and winds rattle seeds."

The Reverend Crawford often wrote labels that were like small poems, kind of haiku-ish. For his specimen of climbing grape, he writes:

> *At edge of old quarry.*
> *Wild grape. Riverbank grape. Frost grape.*

And for a sample of sugar maple, his little label poem reads:

> *Near beech, amongst ironwood, bracken*
> *fern, asters, goldenrod.*
>
> *Sugar maple.*

And, finally, here is his description of St. John's wort:

> *Growing in an abandoned*
> *field, amongst grasses, red*
> *cedars.*
>
> *St. John's wort.*

‹ ‹ ‹ ◊ › › ›

Kingston, Ontario

Acer saccharum Marsh.

MISSOURI: Wayne County. 20 June 1969.
Vicinity of Markham Springs
Recreation Area. Clark National
Forest. West bank of Black River
Ca. 1 mi. n. of Hwy. 49. Sect.
T. 27 N., R. 4 E.

Acer barbatum Michx

Revision by S. Vander Kloet 1969

Evolutionary Logic

specimen

SOUTHERN SUGAR MAPLE
(*Acer barbatum*)
Fowler Herbarium

STRIPED MAPLE
(*Acer pensylvanicum*)
Fowler Herbarium

In the outside world, the trees have begun to turn colour. Inside, I am onto Maples — each specimen taken when the leaves were still green, in spring or early summer, the maple seeds small and barely formed. Some of the maples were collected from golf courses and along the edge of roads. These the collectors called "highway trees."

Because the herbarium is organized according to evolutionary biology, with the oldest-known plants at the beginning of the collection, sometimes there are surprising sequences. After Maples comes Jewelweed (*Impatiens capensis*), the bright orange blooms flattened to brown mush in much the same way that the delicate beauty of an orchid becomes a kind of brown stain on the archival paper.

〈〈〈❍〉〉〉

In 1830, in a chalk pit, in Denbies, Surrey, England, a Miss Parker collected a specimen of the low-growing shrub Aaron's beard (*Hypericum calycinum*) for the herbarium of botanist Sir Charles James Bunbury.

That is almost two hundred years ago and it would be fantastic to find out who Miss Parker was, and whether C.J. Bunbury, like Baron von Mueller, depended on a brigade of women collecting his plants for him, but there is no trail for Miss Parker.

But I do manage to find a woman collector whom I can trace. This is because she used her full name, rather than calling herself "Miss," and because her name was distinctive, and because her descendants were

179

interested in genealogy and kept her story alive and her watercolour paintings of wildflowers in circulation and published her childhood diary.

Lulie Crawford was the daughter of the founder of Steamboat Springs, Colorado, in the Yampa Valley, home of the Ute and Arapaho Peoples. The Crawfords were early settlers and the diary of twelve-year-old Lulie is all about their efforts at subsistence. She details the game they killed and dressed, the planting of vegetables and fruit trees, and gives a blow-by-blow of the weather. She also talks about her greyhound named Legs and seems to be perpetually longing for the mail, which doesn't arrive frequently. I'm sure there were many reasons to delay the mail, but in 1880 the mail courier became lost during a bad snowstorm, emerging from the blizzard three days later with frostbitten feet and his mail bag intact.[3]

When Lulie Crawford was twenty-five, during the spring of 1892, she collected the specimen of dog violet (*Viola riviniana*) that is now in the Fowler Herbarium. And she wrote her full name, in large, flourishing strokes, on the collecting label.

《《《◇》》》

It is rare that an exact location is given for a plant on the old collecting labels (now there are GPS coordinates to pinpoint the precise spot where something was picked), but I find a label from 1897 for a red maple (*Acer rubrum*) that lists the tree as growing in front of a house near where I live, and another dated 1896 that lists Virginia creeper (*Parthenocissus quinquefolia*) climbing up the front of another house, with an exact address given, so I walk over to see if there is anything left of either.

The houses are literally around the corner from one another and in the vicinity of the university, so perhaps the collector lived nearby, or worked at the university and this was his route to work.

There is no red maple outside the house that had one there in 1897, but there are several maples in the same block that are well over a hundred years old, so perhaps the missing tree was of that planting and succumbed to disease, or a storm, or old age.

There is no Virginia creeper on the front of the brick house, but there is a substantial amount of Boston ivy covering the bricks. One kind of creeping plant has been exchanged for another, and I find it pleasing to think of the continuity in this, that this house has probably, for over a hundred years, had a luxurious green vine covering its brickwork, moving slowly upward in the wake of the original Virginia creeper, much the same way mountaineers tread in the steps of those who have gone before them up the mountain face.

‹ ‹ ‹ • › › ›

In the virtual forest of the herbarium, I now find myself in a patch of violets that stretches on and on, file after file, until it is more field than patch.

It is not hard to see the allure of the violet (*Viola*). It has a storied past, features often in literature, and is also associated with soldiers. The Bonapartes chose it for their emblem. And in Mrs. Starr Dana's field guide, she writes, "It seems as if no other flower were so suggestive of the dawning year, so associated with the days when life was full of promise."[4]

Preserving violets as keepsakes has always been popular. They are small and delicate and easy to slip between the pages of a book, or tuck into a letter. The varieties found in North America do not have a scent, so they are not perhaps as romantic a plant as a rose, but their delicacy makes up for their lack of scent, and while not as exotic as orchids, I can see that they have the same kind of magic about them.

One collector has picked different violets from
the small village where he lived, on days near to one
another — May 19 and May 27, in 1913 — a marsh blue
violet (*Viola cucullata*), an alpine violet (*Viola labra-
dorica*), and a downy yellow violet (*Viola pubescens*).
Usually blue flowers keep their colour in death, but
the pale blues of the first two violets have turned to a
cream colour over the hundred years they have been
lying in the file folder.

As it is with all the specimens, each collector
preserves the plant in a different manner. Some of the
violets are slammed down on the page, their stalks and
leaves and flower heads all intermingled, their roots
in a great glob of dirt. It is almost as if the taking of
the violet and its pressing was done in a single action,
yanking it from the forest floor and slapping it down on
the archival paper in one vigorous movement.

Other collectors, or those who end up mounting
their specimens, take great pains to separate out all the
stalks and roots from one another, to point the flower
heads upward towards the viewer, and to create space
between the leaves, so that the violet seems to float on
the page, each part of it distinct and all the connections
shown, each line traceable back to its source.

183

Many of the violet collectors I haven't run across before, and I can find nothing about them. They might have only existed here, among the violets, and not in any other part of this virtual forest. But my idea of violets being similar to orchids is borne out in the fact that one of the violet collectors can also be found in the orchids section.

Abel A. Hunter was an American botanist from Lincoln, Nebraska, who collected orchids in Panama with the great orchid collector Charles Powell, often spending months at a time away from home on expeditions in the early 1900s. His plant for the herbarium, a common blue violet (*Viola sororia*), was collected in his hometown of Lincoln. Unlike the orchid, it is found virtually everywhere, no hunting required.

VIOLET
(*Viola ambigua*)
Fowler Herbarium

These are the places where the various collectors discovered their violets:

"Storm stricken deciduous woodland"

"Luxuriant in rich grass land"

"Flowering near the water; fruiting on sunny cliffs"

"Damp woods"

"edges of granite ledge"

"Wooded hillside"

"Rocky woods"

"Common in mossy swards"

"border of salt meadow"

"Neglected mountain pasture"

The range of habitat shows the hardy, adaptable nature of the plant and I like to picture all of these locales stamped with a bright coin of blue or yellow violet.

‹ ‹ ‹ ◦ › › ›

After Violets come primrose (*Primula vulgaris*), then fireweed (*Chamaenerion*) — noted as sometimes growing around fox dens — then sweet cicely (*Myrrhris odorata*) and giant wild parsnip (*Heracleum mantegazzianum*), and then, weirdly, back to a later primrose, the hairy evening primrose (*Oenothera villosa*).

I'm moving swiftly now, lighting on each new flower and then quickly zipping on, like a bee efficiently visiting each new bloom in its path. What is happening is that I am recognizing a lot of the collectors and it is less likely that I will find new ones as I go along. Sometimes a botanist has a particular liking for one kind of plant — like the man in Violets who picked those around his little village and nowhere else — but more often there are repeats and someone who I have already found in Grasses or Ferns makes another appearance in Primroses or Fireweed.

I've started to realize the limits of the collection the longer I am immersed in it. The limits of this herbarium is that it — like much of anything in North America — is attached firmly to our settler-colonial past. I am trying to unearth stories about the collectors that diverge from this in subtle ways, but it is mostly hard-going and what I feel, now that I am well over halfway through the specimens, is the constricting narrowness of this history. I wonder whether I should abandon my chronological approach in favour of something more spontaneous? What is gained by keeping to the same path, and what is lost? Does it really matter that I look at each and every one of the 144,000 specimens in the Fowler Herbarium?

‹ ‹ ‹ ◦ › › ›

Often with writing, the very thing that is the bright idea at the beginning of a book is the thing that trips you up further in. My idea for this project was to show the interaction between people and nature through time, but this becomes problematic when the people are mostly white colonial settlers. Perhaps I should pay less attention to the collectors and more to the plants themselves?

I am also realizing that the research-heavy nature of this project is forcing me to move at a slower pace than usual, as the rate at which I can move through the file drawers of plants is much slower than the pace at which I can write. So, I find that the research for the book holds the book up. It is as though I am moving at the very pace of the plants themselves, of their growing.

Here is Queen Anne's lace (*Daucus carota*), its 1918 self looking very much like the dried-out version I found in the field this autumn. And here is a later specimen, accompanied by an Ontario government pamphlet[5] from 1946 on the plant — not its charms, but how to eradicate it. The four-page pamphlet presents this plant that I have been so admiring of as a noxious pest and advises that "vigorous action is necessary to suppress Wild Carrot in Ontario." This action is the spraying of a toxic pesticide in the fields and along the roadways, where its appearance has been deemed "unsightly." Like all hate literature, the pamphlet is inflammatory, concluding by saying that Queen Anne's lace "is spreading at an alarming rate and grows more objectionable each year. Every effort should be made to keep it from maturing seed."

Queen Anne's lace is not native to North America: it was brought here by early European settlers as a medicinal herb for the treatment of digestive disorders, kidney and bladder problems, and dropsy.

WILD CARROT
(*Daucus carota*)
Fowler Herbarium

After Queen Anne's lace, there is ground alder or goutweed (*Aegopodium podagraria*), noted on a specimen collected by Henry Eggert in Brooklyn, New York, in 1871 as having possibly come into America with German immigrants.

I move through the dogwoods, noting on one undated specimen from George Clinton that he has used the word "Kinnikinnik" to indicate that the red osier dogwood (*Cornus sericea*) is one of the ingredients in a type of tobacco made and smoked by various Indigenous peoples. Often the inner bark of the dogwood or willow is used, as well as dried sumac leaves, bearberry, arrowroot, mullein, and a variety of other ingredients.

After Dogwoods there are rhododendron (*Rhododendron ferrugineum*) and sheep laurel (*Kalmia angustifolia*), then Labrador tea (*Rhododendron groenlandicum*) and staggerbush (*Lyonia*), trailing arbutus (*Epigaea repens*), creeping strawberry (*Microcachrys tetragona*). Next is alpine bearberry (*Arctous alpine*), then huckleberry (*Vaccinium*), then blueberry (*Cyanococcus*).

‹‹‹ ◦ ›››

With something so common as the blueberry, there are
a multitude of places where it has been collected. On
the many specimens in the herbarium, I take note of
the different descriptions of location. This is a partial
list of where the blueberries were found:

"on an island"
"dry slope"
"open woodland"
"oak-aspen savannah"
"dry ridge under white pine"
"forest of mixed maple-oak conifer"
"in dwarf birch thicket"
"edge of sphagnum bog"
"Blueberry Point"
"rundown pasture land"
"roadsides"
"railroad tracks"
"woods near Fire Tower"
"summit of bluff"
"swamp"
"along old river channels"
"hilltop"
"wet woodland"
"damp thicket"
"barrens"
"sandy ridge"

《 《 《 ◦ 》 》 》

After the blueberries, there is heather (*Calluna vulgaris*), and a sample from 1831 that is still the pale pink colour it would be if it were growing today. After heather, loosestrife (*Lythrum*), then starflower (*Trientalis borealis*) with its small papery, delicate bloom. Then pimpernel (*Anagallis arvensis*), common thrift (*Armeria maritima*), and sea lavender (*Limonium*) — still pale and beautiful in the late 1800s. Black ash (*Fraxinus nigra*) follows white ash (*Fraxinus americana*) and is in turn followed by lilac (*Syringa*).[6]

As I move quickly through the file folders, and through the different dried flowers, I think about what a strange thing it is to have a flower removed so completely from its actual life. There is no sun, no bees, no breeze that moves the blooms, no rain, no crawl of insects, no worms, no wet soil. Each flower isn't accompanied by its brethren, their heads moving together when the wind blows, or drooping collectively from the heat. Each flower is instead pressed flat in an individual file folder, separated from its fellows by a stiff sheath of cardboard.

Centre of facing page:
COMMON LILAC
(*Syringa vulgaris*)
Emily Dickinson

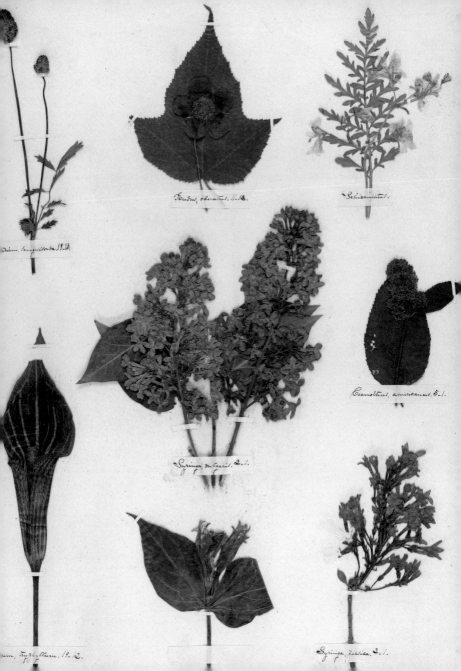

Sium, sanguisorba 19.16.

Rubus odoratus. N.B.

Schizanthus.

Ceanothus americanus. 5.1.

Syringa vulgaris. 2.1.

...um, tryphyllum. 19.12.

Syringa persica. 2.1.

And yet, the flowers still take up their space, still resemble themselves, and I can feel their enduring presence, whereas I am starting to feel like I am the fleeting, temporary thing, that my human life is so brief in comparison to the genetic continuance of this gentian, or this violet. I feel like a small, flickering flame, stuttering along the rows of file cabinets, barely here at all.

This is not something I had anticipated feeling because I had been thinking at the start of this project of how so many plants and animals have been disappearing from our world. I had been thinking of nature as fragile, but now it seems that although humans have the capacity to destroy so much of the natural world, they might actually be one of the most fragile organisms and have the least resilience.

〈 〈 〈 ◦ 〉 〉 〉

Moving this quickly through the plants feels good after all the months of crawling along. I'm now moving at the pace I would be if I were walking through a garden, stopping here and there to admire a particular flower, and then moving on again. I am advancing at what feels like my natural pace now.

Flowers that have commonalities, that resemble one another, or have a relationship to one another are placed in proximity in the herbarium. So, nightshade (*Solanaceae*) follows bittersweet (*Solanum dulcamara*) and precedes nicotiana (*Nicotiana tabacum*) and mullein (*Verbascum*). Digitalis (*Digitalis purpurea*) follows speedwell (*Veronica*) and one of the specimens comes with a note from Hooker, who calls it, "The most stately and beautiful of British herbaceous plants; and one that has obtained great reputation as a medicine."

Some of the common names of the plants, when near to one another, tell a small, cryptic tale — cockscomb (*Celosia*), lousewort (*Pedicularis*), beechdrops (*Epifagus virginiana*), butterworts (*Pinguicula*), bladderworts (*Utricularia*), wild petunia (*Ruellia*), hummingbird plant (*Sesbania grandiflora*).

195

And, when I look through the files, I can see that the blooms of elderberry (Sambucus) look like the viburnum (*Viburnum tinus*) flowers, and that the berry clusters of snowberry (*Symphoricarpos*) resemble the flower clusters of the honeysuckle (*Lonicera*).

《《《◇》》》

It is now the end of October. I have been coming to the herbarium for a year, and I can finally see the end of this work. I am advancing down the last row of metal cabinets and the flowers speed by — fleabane (*Erigeron*), broom (*Cytisus scoparius*), pussytoes (*Antennaria*), pearly everlasting (*Anaphalis margaritacea*), zinnias (*Zinnia elegans*), black-eyed Susans (*Rudbeckia hirta*). Much of what I am looking at in its dried and dead state is still blooming in the fields where I walk in the morning, which makes my inside and outside world collapse a little. The cosmos (*Cosmos bipinnatus*) still weaving in the breeze at the edges of the road are now pressed flat in this file folder and stacked inside this metal cabinet.

And just as though we were moving to the margins of a field, the cabinets reveal burdock (*Arctium*) and thistles (*Cirsium*) after marigolds (*Tagetes*) and sunflowers (*Helianthus*), and finally the vascular plants end with the bright orange disk of hawkweed (*Hieracium*), dried to a burnt umber and flattened in the files.

〈〈〈◊〉〉〉

COMMON SCOTCH THISTLE
(*Onopordum acanthium*)
Fowler Herbarium

Frontenac Co.

Name Cnicus lanceolatus, Hc

Delesseria sinuosa

Algae & Fungi

specimen

ALGAE
(*Delesseria sinuosa*)
Fowler Herbarium

On my last day at the herbarium, on the drive up, I meet a young fox in the middle of the road. She is casually grooming herself and continues this task when I stop the car to let her pass. She looks me square in the eyes and ambles slowly off into the leaf-strewn woods, without the slightest flicker of fear at my presence. It seems fortuitous to end my journey with a fox, as I began it also with a fox, and there is a serendipity to the two sightings, almost a full year apart. The young fox is a juvenile, so they are probably not the same fox, but I like the synchronicity all the same.

《 《 《 ◊ 》 》

After vascular plants, there are still a few more files to look through. These are Algae and Fungi. There are also lichens and mosses in the Fowler collection, but they haven't been catalogued yet, and so I decide to end with algae. The fungi specimens are either boxed or packed into paper envelopes and I don't want to damage them by undoing their packaging to examine them. The mushrooms that have been glued to archival sheets, I notice have hardened to the consistency of wood.

The algae are otherworldly, beautiful and delicate, with most of them retaining vivid colour. Many look like ornate lace, or the whispery tendrils of root systems. They are a dreamy departure from the vascular system of stems and leaves that I have become so accustomed to.

While many of the collectors have repeated in the last dozen cabinets of the vascular plants, I find different people in Algae. Chief among them is Josephine Elizabeth Tilden, an expert in Pacific algae in the early twentieth century. Employed by the University of Minnesota as its first woman scientist, she travelled by canoe along the coast of British Columbia in 1900, finding a deserted stretch upon which to build a research station and host, every summer, students and visiting professors to study geology, algology, zoology, taxonomy, and lichenology. In the evenings the participants put on plays and told stories. The research station was funded by Tilden's own money, and she tried, to no avail, to get the university to take over the costs. In 1907 the station was shut down when Tilden could no longer afford to run it, but she continued to lead collecting expeditions all over the world. The old Tilden station site is now inside Juan de Fuca Provincial Park.

Ethel Sarel Barton was a specimen collector for the British Museum. In 1900 she began reorganizing the genus *Halimeda*, reducing it to seven species from over two dozen. At the time this was an important contribution to phycology. In some of the algae specimens, Tilden has asked Barton to verify a particular specimen, so she obviously had respect for her knowledge. Also, interesting that the women, who would have been heavily outnumbered by their male counterparts, were in communication with one another. Barton also worked with the Dutch phycologist Anna Weber-van Bosse, on another collection of *Halimeda* specimens.

There seem to be more women in the algae field than in vascular plants, and I wonder if this was in part because it was easier for a woman to walk alone along the seashore than to wander alone through the woods and forests? Also, seaweed collecting, like ferning, was a popular Victorian pastime. Queen Victoria herself was a devotee, as was the writer George Eliot.

One of the most accomplished seaweed "hunters" was the British woman Margaret Gatty, who began writing her definitive guide on the subject, with

hand-drawn illustrations, in 1848. *British Sea-Weeds* was published in 1872. In the book, Gatty advised women collectors to wear men's boots for the task, for their stability. "*Feel* all the luxury of not having to be afraid of your boots," she wrote. And when trying to persuade her sister-in-law to take up the hobby, she reminded her of the tedium of the domestic life for women, and advised that, "Your seaweed hours will be a sort of necessary repast for you."

STRINGY ACID KELP   (*Desmarestia viridis*) Fowler Herbarium

Dasya elegans. Ag.

Wood's Holl. Mass. 1894.

Richard Lees

I think of Margaret Gatty, tromping up and down Britain's southeast coast in men's boots and women's petticoats, taking her seaweed specimens home to draw and catalogue them. A whole lifetime spent doing this, as the other collectors in this herbarium spent their lives in the same way. It suddenly seems to me like a devotional exercise. And I feel joined across time to these collectors because this is also what my work in the herbarium, and my drawing practice, has been during this year — an act of devotion, the work of common prayer.

‹ ‹ ‹ ◦ › › ›

Leaving the building and walking through the woods to my car, I notice the changes in the giant puffball mushroom I have been observing since early in the fall. It no longer glows white, but has deflated like old leather, sunken and mouldy — a slow, darkening collapse. When prodded, brown dust rises from its sag. The mushroom's flesh is turning to powder. Soon there will only be a dark, rubbery snout, poking through the ice among the winter trees.

RED ALGAE
(*Dasya elegans*)
Fowler Herbarium

‹ ‹ ‹ ‹ ‹ ● › › › › ›

On Thoreau

specimen

PINK TICKSEED
(*Coreopsis rosea*)
Henry David Thoreau

It would be remiss not to talk about Henry David Thoreau or his herbarium in this book about plants, in both their live and pressed forms. For North American settlers, Thoreau is probably the foremost example of someone who lived close to the earth and documented his experiences in nature. His way of thinking about nature has profoundly influenced how we think about it today.

Thoreau's plan to "live deliberately" was about intimately observing the natural world around him, and in his relatively short life (he died at the age of forty-four of tuberculosis) he wrote a lengthy daily journal, as well as a handful of books, including the famous treatise *Walden*, about his time spent in a tiny cabin on the edge of Walden Pond near Concord, Massachusetts.

In the way we tend to romanticize the past, it is easy to think that Thoreau enjoyed a greater bounty of nature than we do now, and that our present-day experience of the natural world is paltry in comparison. But that isn't true. In Thoreau's nineteenth-century environment much of the forests around him were being chopped down by the logging industry and the wildlife was consistently over-hunted to the point of near extinction for some species. (Thoreau only ever saw a couple of deer in his lifetime and considered it a momentous event.)[7] Also, the cinders from the wood-fired steam trains were constantly setting the forest on fire, so there was an apocalyptic cast to Thoreau's world, which added perhaps to the urgency he felt at trying to write down what he saw and experienced.

As well as documenting the natural world in writing, Thoreau also made a herbarium that, at the time of his death in 1862, numbered over nine hundred specimens. The plants were those he had found in his Concord environs and that he was interested in identifying. Instead of a collecting bag, he carried the specimens home tucked under his straw hat.

Perhaps this precarious method of transporting his specimens is why he brought home multiple examples of the same plant. His sheets are jammed full of as many dried plants as can fit on a single sheet of paper.

Thoreau was assisted with his herbarium by his sister Sophia, who also collected for her own herbarium. She helped press and mount her brother's specimens as well as her own, so when I look at Thoreau's sheets, I might be looking at his handiwork in laying them out, or I might be seeing his sister's decisions. Sophia was a talented artist and there is an aesthetic concern to the placing of the specimens that suggests they might well have been arranged by her. Sophia was more inventive in her botanizing. She once made a checkerboard that contained small plant specimens, mostly ferns, and she inscribed one of Henry's poems onto a set of five shagbark hickory leaves.

Whoever glued the pressed plants onto the paper was probably left-handed, as the plants mostly list to the right, as though a prevailing wind had blown through all the pages.

One page is full of coloured fall leaves and another is stuffed so full of lily pads that they appear to be floating on the surface of the paper as though it were water.

There is a water crowfoot (*Ranunculus aquatilis*) that loops around like stitching on the hem of a dress, and a climbing fumitory (*Adlumia cirrhosa*) that looks like an embroidered letter S. One page is full of different kinds of violets, as thick on the paper as they would have been in the wood where they were found. There is thyme (*Thymus*) in sprigs that seem laid out as though on a chopping board to be used in cooking, and a specimen of yellow wild indigo (*Baptisia tinctoria*) that looks like it's swimming — each strand an arcing line reaching across the surface of the paper.

Whoever put the multiples of chickweed (*Stellaria media*) on the page arranged them in a star-shaped pattern to resemble the individual flower, and a series of catkins were placed overtop one another, so that they look like a waterfall. Oak leaves of different sizes (collected in the autumn when they were bronze and brown) are overlapped on top of one another the way they would be on the forest floor. Grasses bend the way they would if a wind had stirred them. A clump of pine boughs has a tiny pine cone, like a small jewel studded in the centre of the dried needles.

The sensibility of the person who set out the herbarium was a fine one, and it is clear that they considered the page where the plants were attached to be similar to a blank canvas, that the plants were to be fashioned into an image that would have visual appeal. It makes me realize that there could be great range in herbaria, depending on whom the collection of dried plants was being made for, and what were the guiding principles (whether artistic or scientific) of the person who was displaying the plant specimens.

Thoreau had a relationship with the plants before he preserved them. He liked to think of them as individuals who shared the same world as he did, as fellow

211

travellers, and he was making a "Kalendar" of Concord for what bloomed when and where, and the various states of the various flowers throughout their lives. He noted, "I am interested in each contemporary plant in my vicinity — and have attained to a certain acquaintance with the larger ones. They are cohabitants with me of this part of the planet, and they bear familiar names. Yet how essentially wild they are — as wild really as those strange fossil plants whose impressions I see on my coal." [8]

Thoreau's musings on the plants around him were both botanical and poetic. He says of a basswood tree in 1852: "The branches of this tree touch the ground, and it has somewhat the appearance of being weighted down with flowers. The air is full of sweetness. The tree is full of poetry." [9]

He noted that hawkweed was a sort of "autumnal dandelion" and that the flowers found in the fields of August and September resembled the sun — sunflowers, asters, and goldenrod. Of chicory he wrote, "This weed is handsomer than most garden flowers." [10]

Thoreau rambled the woods and swamps and fields, taking note of the location of his floral companions and remarking when they had come into bloom and for how long. He noted when the perfume of the fragrant flowers

was its strongest, and also noted the smell of those flowers that didn't have a strong fragrance. It is touching to think of him on his hands and knees in the forest, sniffing a violet or columbine.

He had hoped for a long life, in order to get all the botanical cataloguing done, to get his Kalendar finished, but he was dead right in the middle of it all, at forty-four.

"It will take you half a lifetime," he wrote, "to find out where to look for the earliest flower."[11]

‹ ‹ ‹ › › ›

ROSE (*Rosa lucida*) Henry David Thoreau

Epilogue

BEACH-HEAD IRIS
(*Iris hookeri*)
Fowler Herbarium

The observations that I have made of the natural world
last in my mind because they were hard won. They were
gained by hours and hours of watching or walking, hours
and hours of looking but not entirely seeing, until the
moment when some new piece of knowledge swam into
consciousness. These moments of clarity are perhaps one
of the greatest pleasures of being a sentient animal.

For hours once, I walked through rainy countryside,
watching my feet the whole time because the path I was
on was slippery and covered with tree roots. At the end
of the day, I realized that, in looking at all those tiny
flowers along the route, I had discovered a fundamental
truth about them — flowers close to rain as they close to
darkness.

Being at the herbarium, in the presence of the dead
plants, has made me more cognizant of the natural world
around me — even more so than before. My observations
have become honed and specific. When the lake freezes
shut in the winter, I watch as the water solidifies and
can see how it attaches to itself in a frozen state by a
series of ice feathers that slowly harden and gloss over.

Driving along the highway that borders the marsh
near the city, I suddenly see how the marsh is the hinge
between lake and land, where blackbirds sway on rushes,

and herons rise on stiffened wings. Where water is a form of darkness, and the choir of wild iris sings with meadowsweet and willow. It is neither solid ground, nor entirely melt, but shifts its state to what is found, matching creature and season. Giving us, too, relief from absolutes, a fate where we can dream ourselves as sway or rise, or earthly song.

When I started this book, I had the idea that the herbarium was a kind of wilderness and I would enter it and see what I could find there, as though I were entering an actual wilderness, and the journey through the dried plants would yield some answers to my questions about humans and nature. But now at the end I realize that it turned out backwards. The herbarium was the steady thing, the constant thing, and it was my life that was the wilderness. During the year that I was looking through the plant specimens, I was, many times, in a kind of emotional wilderness — feeling the aftershocks of a decade of human deaths, and then experiencing the death of my dog while working on the book. One of the things that kept me grounded during this time was my exploration of the herbarium, the reassuring routine of looking through file after file of plant specimens. The process was comforting, reliable, at a time when I was feeling, more often than not, sad and adrift. How strange, and not strange, that things should turn out this way!

Through my time at the herbarium, and in the writing of this book, I have been looking for a way to talk about the present place and time, of where we are in terms of nature on this planet, about how to feel optimism in the face of mass species extinction and climate crisis. But what I have realized is that life itself is an optimistic process. It is about regeneration and forward movement, about growth and change. The imperative of life is an imperative of optimism, and I don't think this inherent spirit of optimism needs to be subverted because life on Earth is in a very grim period. We can't merely focus on what is gone or disappearing. We also need to pay attention to what continues. It is a balance — one that is hard to maintain as it is easy to fall into despair about our current age — but a balance that, I feel, it is necessary to maintain, and that we need to work hard to maintain, and not tip over into despair.

There is much to be joyful about in this world, and we need to feel a certain amount of optimism or joy to do what needs to be done to effect change. Despair is not a great motivator, but hope is.

So, while I could end this book at the herbarium, or in the fields where I walk every morning, I have decided to end with a story, a true story about another time. This story contains both the hope and despair of what it means to live on this Earth in this moment in time, and yet the story takes place almost two thousand years ago.

〈〈〈◇〉〉〉

The Roman Colosseum was completed in 80 AD and used as a bloody amphitheatre for four centuries. Over 400,000 people were killed there and over a million animals were slaughtered. Certain species of animals were severely depleted or made extinct from this carnage. The North African elephant was wiped out, as was the Nile hippo. It was a horrific time for the wild creatures of the world.

Animals were captured and transported to Rome from all around the globe. Tigers were caught in Armenia, boars and bears from northern Europe, and elephants, lions, ostriches, leopards, and hippos

Drawing from Richard Deakin's *Flora of the Colosseum of Rome.*

Interior of Colosseum   1853

were dispatched from northern Africa. The overland transport through Europe was slow, but a ship loaded with wild animals from Africa could make it back to Italy in two days. [1]

Some animals died in transport — from stress, or starvation, or disease — and the ones that survived the journey were kept in pens in the basement of the Colosseum until the day of their execution.

After its four hundred years of active use, the Colosseum fell into disrepair. The city had been sacked numerous times, and half of its population had fled. Also, there were multiple earthquakes that had damaged the structure of the building. In medieval times, houses and churches were built inside the ruins, and around 1200 it was occupied by various artisans — blacksmiths, cobblers, cart makers — who lived in these houses and planted orchards and gardens. Building materials from the Colosseum were plundered and in later centuries the ruins became the dwelling place for criminals and homeless people. In the 1700s and 1800s there was more damage to the building from earthquakes, and in 1870 the debris and vegetation was removed and restoration work was begun that continues to this day.

In the mid-1800s, a British doctor named Richard
Deakin went to Rome to work there for a couple of
years. As he was not only a doctor but also a very good
amateur botanist — he had previously published the
first affordable flora guide to Britain, the *Florigraphia
Britannica* (1837–1848) in four volumes, with all the
illustrative material drawn himself — he decided to do
a botanical inventory of the Roman Colosseum. The
building was an overgrown garden and Richard Deakin
found 420 different species of plants growing in and
around its six-acre circumference. He published his
findings in a little book, illustrated with his own water-
colours, called *Flora of the Colosseum of Rome* (1855).

What was interesting about Richard Deakin's
inventory was that almost a fifth of the plant species
that Deakin discovered were not native to Italy. He put
forward the theory that the seeds for these plants had
travelled to Rome in the digestive tracts and on the fur
and skins of the animals who had been killed for sport
in the amphitheatre.

While some of these plants might have been brought
in by some of the human occupants in the years since it
was used for blood sports, most of the plants were from
northern Africa, which is where most of the animals
had come from.

These plants included many grasses and grains —
bentgrass, bufflegrass, and wood barley — and weeds
such as dead-nettle and sandspurry. They included
edible plants such as wild celery and garden cress,
Indian dill and edible burdock. And they also numbered
many flowers — bellflowers, flax, larkspur, common
stork's-bill, and wild madder.[2] Many of these plants
were native to North Africa and all of them can be
found growing in Italy today, centuries and centuries
after the last animal was slaughtered in the Colosseum.

Connecting the flora to the fauna joins the living
to the dead and makes a bridge between them.

So, here is the African lion, pacing in his cage in
the stone labyrinth below the Colosseum, where above
fifty thousand spectators eagerly await his slaughter.
He moves anxiously in his small pen, the sides of his
skin touching the walls at each turn.

Here is the lion, far from home and sanctuary and
safety, from his wild and true life, and mere hours
away from his death. He paces and turns and brushes
the walls with his flanks, until a tiny seed of oat grass
rubs off his coat, nestles into a crack in the stone,
slowly unfolds, and begins to grow.

‹ ‹ ‹ ◊ › › ›

NOTES

WINTER

1 Field notebook of Roland Beschel, New Brunswick, 1957. Collection of Fowler Herbarium.

2 *Ibid.*

3 Field notebooks of Roland Beschel in the collection of the Fowler Herbarium. All following quotations from Beschel's notebooks are from this source.

4 Vascular plants are land plants that contain vascular tissues that move minerals and water through the plant, as well as conducting photosynthesis.

5 A. Crowder, "The collection of bryophytes in the Fowler Herbarium, Queen's University, Kingston, Ontario." *Canadian Field Naturalist* 88 (1974): 47–55.

6 Deb Metsger, curator of the ROM herbarium, in conversation with the author.

7 Letter from John Keats to his brother, Tom, on June 27, 1818. Published in *Selected Letters by John Keats* (Penguin Classics, 2015).

8 Edward Joseph Lowe, *A Natural History of New and Rare Ferns* (Groombridge and Sons, 1865).

9 Penny Olsen, *Collecting Ladies: Ferdinand Von Mueller and Women Botanical Artists* (National Library of Australia, 2013).

10 *Ibid.*, 10.

11 William Harrison Taylor, Peter C. Messer, *Faith and Slavery in the Presbyterian Diaspora* (Rowman & Littlefield, 2016).

12 Diary of William Norman Rudolf, William Norman Rudolf fonds, Nova Scotia Archives.

13 John Inglis, *In the New Hebrides: Reminiscences of Missionary Life and Work, Especially on the Island of Aneityum, from 1850 Till 1877* (T. Nelson and Sons, 1887).

14 Mary Treat, *Home Studies in Nature* (New York: American Book Company, 1885).

15 Historical weather data for July 16, 1903, Environment and Climate Change Canada, https://climate.weather.gc.ca/historical_data/search_historic_data_e.html.

16 Edith Bolan Ogden, *The Ferns of Maine* (Orono, Maine: University Press, 1948).

17 Adele Crowder and Vivien Taylor, "Mrs. Traill, Mrs. Roy and Miss Boyd: Plant Collectors in 19th Century Upper Canada," *Occasional Papers Series No. 1* (W.D. Jordan Library, 2003).

18 Advertisement in *Hardwicke's science-gossip: an illustrated medium of interchange and gossip for students and lovers of nature* (Robert Hardwicke) (London: 1887).

19 *Ibid.*, 1888.

20 Rhoda M. Love, "Wilhelm Nikolaus Suksdorf (1850–1932), Pioneer Botanist of the Pacific Northwest," *Pacific Northwest Quarterly* (Fall 1998).

21 William A. Weber, "The Botanical Collections of Wilhelm N. Suksdorf 1850–1932," *Research Studies, State College of Washington*, vol. 12 (1944).

22 Botanists are known as "splitters" or "lumpers" — those, like Suksdorf, who want to acknowledge the differences within an individual species by subdividing it, and those who want to throw all the variations together under the same categorization.

23 W.S.M. D'Urban, "Catalogue of Plants Collected in the Counties of Argenteuil and Ottawa, in 1858," *The Canadian Naturalist and Quarterly Journal of Science with the Proceedings of the Natural History Society of Montreal*, vol. 6 (1861).

24 *Research Report — The Location of the Northern Boundary, Mississagi River Indian Reserve #8, at Blind River*, Indian and Northern Affairs (Government of Canada, 1980).

25 W.S.M. D'Urban, "Catalogue of Plants Collected in the Counties of Argenteuil and Ottawa, in 1858," *The Canadian Naturalist and Quarterly Journal of Science with the Proceedings of the Natural History Society of Montreal*, vol. 6 (1861).

SPRING

1 Walter Deane, "Michael Schuck Bebb," *Botanical Gazette* no. 2 (February 1896): 53–66.

2 *Ibid.*

3 Cyrus Pringle and Rufus M. Jones, *The Record of a Quaker Conscience* (New York: Macmillan, 1918).

4   Walter Arthur Donnelly, *The University of Michigan, an Encyclopedic Survey* (UM Libraries Michigan, 1958).

5   G.P. Walsh, "Squire, James," *Australian Dictionary of Biography*, vol. 2 (Melbourne University Press, 1967).

6   Interview by the author with Nora Talkington, curator of Navajo Nation Herbarium (November 2019).

7   "Botany Is Their Hobby," *San Francisco Chronicle* (April 22, 1894).

SUMMER

1   Mrs. William Starr Dana, *How to Know the Wildflowers* (New York: Charles Scribner's Sons, 1894).

2   *Ibid.*, 90–91.

3   Francis Halle, *Atlas of Poetic Botany* (Cambridge: MIT Press, 2018).

4   A system of classification invented by Swedish botanist Carl Linnaeus (1707–1778) that is based on a taxonomical hierarchy of nature.

5   M.L. Fernald was a noted American botanist. He collected his specimen of purging flax in Sydney, Nova Scotia, on August 20, 1902, and noted that he found it "Abundant in damp spots in an old field."

6   Erich Haber, "In Memoriam — John 'Jack' Montague Gillett, 1918–2014," *Botanical Electronic News*, issue 491 (April 23, 2015).

AUTUMN

1   William H. King, "Marcus E. Jones (1852–1934)," Newsletter of the Utah Native Plant Society, vol. 34, no. 6 (November 2011).

2   Proceedings of the Entomological Society of Ontario 1896, 39.

3   Lulita Crawford Pritchett, ed., *Diary of Lulie Margie Crawford*, 1985.

4   Mrs. William Starr Dana, *How to Know the Wildflowers* (New York: Charles Scribner's Sons, 1894), 230.

5   "Wild Carrot" (Toronto: Ontario Department of Agriculture, May 1946).

6   After gentians, there is common periwinkle (*Vinca minor*), then dogbane (*Apocynum cannabinum*), a poisonous plant (part of its Latin name, *Apocynum*, means "poisonous to dogs"). Following dogbane is milkweed, with its silken strands still soft after decades and decades of being kept in the file folder. Next is nettles (*Urtica dioica*), and then convolvulus, or hedge bindweed (*Calystegia sepium*), described rather poetically on one label from Alberta in 1961 as "Climbing, in dense undergrowth."

    Morning glory (*Ipomoea*), the colourful version of hedge bindweed, are next. And then there are phlox (*Phlox paniculata*), gilia (*Aliciella latifolia*), waterleaf (*Talinum fruticosum*), phacelia (*Phacelia tanaceti-folia*), heliotrope (*Heliotropium*), and comfrey (*Symphytum*). Following behind them is common hounds-tooth (*Cynoglossum officinale*), and a label from Elgin, Ontario, in 1953 that describes the plant as having a "mousey odor," which seems an odd way to describe something and suggests the collector had an overly familiar association with rodents.

Fiddlenecks (*Amsinckia*) are next, then borage (*Borago officinalis*), and forget-me-nots (*Myosotis*). One of the specimens of field gromwell (*Lithospermum arvense*) comes from the botanical garden at Queen's University. Its seeds were once ground and used as a treatment for arthritis, while the leaves were infused and used as a sedative.

After this comes bugloss (*Echium vulgare*), also known as blueweed, blue vervain (*Verbena hastata*), rosemary (*Salvia rosmarinus*), scull-cap (*Scutellaria*), anise hyssop (*Agastache foeniculum*), catnip (*Nepeta cataria*), dragonhead (*Dracocephalum*), self-heal (*Prunella vulgaris*), hemp nettle (*Galeopsis tetrahit*), dead-nettles (*Lamium*), motherwort (*Leonurus cardiaca*), rough hedge nettle (*Stachys rigida*), salvia (*Salvia officinalis*), beebalm (*Monarda*), pennyroyal (*Mentha pulegium*), calamintha (*Calamintha nepeta*), hyssop (*Hyssopus officinalis*), and mountain mint (*Pycnanthemum*).

7 Laura Dassow Walls, *Henry David Thoreau: A Life* (Chicago: University of Chicago Press, 2017).

8 Henry David Thoreau's journal entry for June 5, 1857. Damion Searls, ed., *The Journal of Henry David Thoreau 1837–1861* (New York: NYRB Classics, 2009).

9 *Ibid.*, July 16, 1852.

10 *Ibid.*, July 9, 1851.

11 *Ibid.*, April 2, 1856.

EPILOGUE

1 Michael MacKinnon, "Supplying Exotic Animals for the Roman Amphi-theatre Games: New Reconstructions Combining Archaelogical, Ancient Textual, Historical and Ethnographic Data," *Mouseion*, series III, vol. 6, no. 2 (2006): 137–161.

2 Richard Deakin, *Flora of the Colosseum of Rome* (London: Groombridge and Sons, 1855).

ACKNOWLEDGEMENTS

Gratitude and thanks, as always, to my agent, Clare Alexander.

Thank you to my editor, Susan Renouf, for guiding this project from beginning to end.

Thanks to Shannon Parr, Jessica Albert, and everyone at ECW Press.

This book could not have been written without the cooperation of the Queen's University Biological Station (QUBS). I thank all those associated with QUBS — in particular, Steve Lougheed, Sonia Nobrega, and Aaron Zolderdo.

Adele Crowder was generous with her memories and time, for which I am very grateful.

Thank you to Deb Metsger.

Adriana Lopez Villalobos helped me with this project in so many ways. I am grateful for her generosity and kindness and her passionate dedication to the collection of plants in the Fowler Herbarium. Thank you so much.

Thanks to Laura Jean Cameron and Heather Home.

Thanks to my erstwhile walking companion, Kirsteen MacLeod. I couldn't have done this without you.

Thank you Mary Louise Adams, Eleanor MacDonald, Susan Mockler, Pamela Mulloy, Susan Olding, Marco Reiter, Su Rynard, Sarah Tsiang. Thanks, Charlotte. You were the best.

An excerpt of this book appeared in *The New Quarterly* issue 152, fall 2019.

Thanks to the Ontario Arts Council for funding during the writing of this book.

Li-Young Lee, excerpt "From Blossoms" from *Rose*. Copyright 1986 by Li-Young Lee. Reprinted with the permission of The Permissions Company, LLC on behalf of BOA Editions Ltd., www.boaeditions.org.

Thank you, Nancy, for keeping the home fires burning.

SELECTED BIBLIOGRAPHY

Atlas of Poetic Botany by Francis Halle; MIT Press; Boston; 2018.

The Botanizers: Amateur Scientists in Nineteenth-Century America; The University of North Carolina Press; Chapel Hill, NC; 2011

Both the Emily Dickinson Herbarium and Thoreau's Herbarium are available to view online through Harvard University Herbaria and Libraries Digital Collections.

Braiding Sweetgrass: Indigenous Wisdom, Scientific Knowledge and the Teachings of Plants by Robin Wall Kimmerer; Milkweed Editions; Minneapolis, MN; 2015.

Flora of the Colosseum of Rome by Richard Deakin; Groombridge and Sons; London; 1855

Grasses of Ontario by William G. Dore; Agriculture Canada; Ottawa; 1980.

Herbarium: The Quest to Preserve and Classify the World's Plants by Barbara M. Thiers; Timber Press; Portland, OR; 2020

Home Studies in Nature by Mary Treat. First published in 1885 by Harper & Brothers, New York, but now widely available as a reprint.

How to Know the Wildflowers by Mrs. William Starr Dana; Charles Scribner & Sons; Austell, GA; 1893.

The Journal of Henry David Thoreau 1837–1861; NYRB Classics; New York; 2009

A Manual of Botany of the Northern United States by Asa Gray, first published in Boston in 1848 by J. Munroe, but with many subsequent editions.

Philosophia Botanica by Carl Linnaeus — first published in 1751, but now available from Oxford University Press; London; 2006.

IMAGE CREDITS

Published by ECW Press
665 Gerrard Street East   Toronto, Ontario   M4M 1Y2
416-694-3348 / info@ecwpress.com

Editor for the Press: Susan Renouf
Book design: Natalie Olsen
Cover image: Dickinson, Emily, 1830–1886. Herbarium, circa 1839–1846. 1 volume
(66 pages) in green cloth case; 37 cm. MS Am 1118.11, Houghton Library ©
President and Fellows of Harvard College. Seq. 19.
Author photo: Ayelet Tsabari

Library and Archives Canada Cataloguing in Publication
Title: Field study : meditations on a year at the herbarium / Helen Humphreys.
Names: Humphreys, Helen, 1961– author.
Identifiers: Canadiana (print) 20210210958 | Canadiana (ebook) 20210211296 |
ISBN 978-1-77041-534-8 (hardcover) | ISBN 978-1-77305-776-7 (EPUB) |
ISBN 978-1-77305-777-4 (PDF) | ISBN 978-1-77305-778-1 (Kindle)
Subjects: LCSH: Herbaria. | LCSH: Herbaria—History. | LCSH: Herbaria—Popular
works. | LCSH: Botany—Popular works.
Classification: LCC QK75 .H86 2021 | DDC 580.74—dc23

The publication of Field Study has been generously supported by the Canada
Council for the Arts and is funded in part by the Government of Canada. Nous
remercions le Conseil des arts du Canada de son soutien. Ce livre est financé
en partie par le gouvernement du Canada. We acknowledge the support of the
Ontario Arts Council (OAC), an agency of the Government of Ontario, which last
year funded 1,965 individual artists and 1,152 organizations in 197 communities
across Ontario for a total of $51.9 million. We also acknowledge the contribution
of the Government of Ontario through the Ontario Book Publishing Tax Credit,
and through Ontario Creates for the marketing of this book.

Canada Council   Conseil des arts
for the Arts   du Canada

Canada

ONTARIO
CREATES

ONTARIO ARTS COUNCIL
CONSEIL DES ARTS DE L'ONTARIO
an Ontario government agency
un organisme du gouvernement de l'Ontario

MIX
Paper from
responsible sources
FSC
www.fsc.org   FSC® C016245

Printed and bound in Canada
Printing: Friesens   5 4 3 2 1